Fauna Entomologica Scandinavica

Volume 16   1986

The Saltatoria

(Bush-crickets, crickets and grasshoppers)

# of Northern Europe

by

Knud Th. Holst

E. J. Brill/Scandinavian Science Press Ltd.

Leiden · Copenhagen

Printed by
Vinderup Bogtrykkeri A/S
7830 Vinderup, Denmark

ISBN 90 04 07860 6
ISBN 87-87491-27-3
ISSN 0106-8377

Authors address:
Lindevej 3B, Hareskov
3500 Værløse, Denmark

SWEDEN

Sk.	Skåne	Vrm.	Värmland
Bl.	Blekinge	Dlr.	Dalarna
Hall.	Halland	Gstr.	Gästrikland
Sm.	Småland	Hls.	Hälsingland
Öl.	Öland	Med.	Medelpad
Gtl.	Gotland	Hrj.	Härjedalen
G. Sand.	Gotska Sandön	Jmt.	Jämtland
Ög.	Östergötland	Ång.	Ångermanland
Vg.	Västergötland	Vb.	Västerbotten
Boh.	Bohuslän	Nb.	Norrbotten
Dlsl.	Dalsland	Äs. Lpm.	Äsele Lappmark
Nrk.	Närke	Ly. Lpm.	Lycksele Lappmark
Sdm.	Södermanland	P. Lpm.	Pite Lappmark
Upl.	Uppland	Lu. Lpm.	Lule Lappmark
Vstm.	Västmanland	T. Lpm.	Torne Lappmark

NORWAY

Ø	Østfold	HO	Hordaland
AK	Akershus	SF	Sogn og Fjordane
HE	Hedmark	MR	Møre og Romsdal
O	Opland	ST	Sør-Trøndelag
B	Buskerud	NT	Nord-Trøndelag
VE	Vestfold	Ns	southern Nordland
TE	Telemark	Nn	northern Nordland
AA	Aust-Agder	TR	Troms
VA	Vest-Agder	F	Finnmark
R	Rogaland		

n northern   s southern   ø eastern   v western   y outer   i inner

FINLAND

Al	Alandia	Kb	Karelia borealis
Ab	Regio aboensis	Om	Ostrobottnia media
N	Nylandia	Ok	Ostrobottnia kajanensis
Ka	Karelia australis	ObS	Ostrobottnia borealis, S part
St	Satakunta	ObN	Ostrobottnia borealis, N part
Ta	Tavastia australis	Ks	Kuusamo
Sa	Savonia australis	LkW	Lapponia kemensis, W part
Oa	Ostrobottnia australis	LkE	Lapponia kemensis, E part
Tb	Tavastia borealis	Li	Lapponia inarensis
Sb	Savonia borealis	Le	Lapponia enontekiensis

USSR

Vib   Regio Viburgensis   Kr   Karelia rossica   Lr   Lapponia rossica

Contents

Introduction

The present volume of "Fauna Entomologica Scandinavica" treats the 45 native species of Saltatoria known from Northern Europe, i.e. Schleswig-Holstein, Denmark, Norway, Sweden, Finland and the adjacent Soviet areas of Eastern Fennoscandia. Also 2 introduced species and one accidental visitor are included. The volume contains keys to the systematic categories, descriptions of all taxa, and information on the local and total distribution, their biology, and stridulation. The text of this book contains more biological information, including a description of sound production, than is found in most other handbooks.

Acknowledgements

The collections of Saltatoria in the Zoological Museum, Copenhagen, and the Museum of Natural History, Århus, were identified in connection with the preparation of Vol. 79 of "Danmarks Fauna" (Holst, 1970). Since then, the author has worked on several other Scandinavian collections (including Finnish ones), and the author wishes to thank the following persons for the loan of material and for information; Lita Greve Jensen and Astrid Løken, Zoologisk Museum, Bergen; Albert Lillehammer, Zoologisk Museum, Oslo; Henrik W. Waldén, Naturhistoriska museet, Gothenburg; Roy Danielsson, Entomologiska museet, Lund; Lars Wallin, Zoologiska museet, Uppsala, and Anders Albrecht, Universitetets zoologiska museum, Helsinki.

The author has travelled to various parts of Denmark and Holstein and also to Southern Sweden (Skåne, Blekinge and Öland) and Southern Norway (Agder, Telemark, Akershus, Opland and Hedmark). Specimens were collected on these journeys, and sound production was recorded on a Uher 4000 Report IC. Thanks are due to the Danish Natural Science Research Council, who financed these journeys and the translation of the manuscript, and to the Danish Ministry of Education for granting leave of absence.

A very special word of thanks is due to S. Boel Pedersen, Acoustic Laboratory, Technical University, Lyngby, Denmark for his invaluable assistance in the production of oscillograms, and to D. Chr. Clayre (M. Sc., Lond.) for his careful execution of the translation work.

Only the 45 species found in natural surroundings are detailed in the list of distribution set out in the final section of the book.

Literature

About half the species of Saltatoria found in Fennoscandia and Denmark were described by Linnaeus (1758, 1761, 1767). *Bryoderma tuberculata,* which was once very common on the moorlands of Jutland, was described by Fabricius (1775) in his Systema entomologiae. One of the first volumes of "Danmarks Fauna" (No. 6) was on the Orthoptera, and was written by Esben-Petersen (1909). Knaben (1943) reviewed the Orthoptera of Norway; Ander (1945, 1953) published a catalogue of the Orthoptera of Sweden; Holst (1970) published a new edition of "Danmarks Fauna"; Albrecht (1979) published a catalogue of the Orthoptera of Finland, and Wallin (1979) released a cassette of sound production in the Saltatoria of Sweden.

Thus the Orthoptera of the region may be said to be quite well known.

Characteristics

The Saltatoria of Northern Europe are largely medium-sized to big insects with compressed or cylindrical bodies and enlarged hind legs for jumping. They are thus not only always easy to identify on the basis of their appearance but also from the way they move. There are, however, exceptions to the general plan, one example of such being *Gryllotalpa gryllotalpa*.

The head is comparatively large, jointed at right angles to the body, the mouth thus being directed downward. All three thoracic segments are large. The prothorax is freely

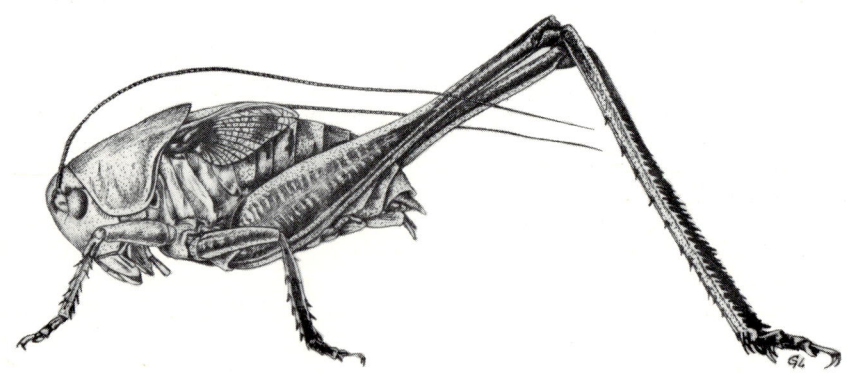

Fig. 1. Last instar nymph of *Decticus verrucivorus* (L.) ♂.
Length about 20 mm.

movable and always markedly larger than the meso- and metathorax, which are fused. The pronotum is bent downwards into two paranota, giving it a saddle-shaped form. The paranota are separated from the superior surface of the pronotum either by a round or sharp edge. The hind legs are strong, and enlarged for jumping. The coxae are free and widely separated. The tarsus consists of three or four segments. The costal vein of the fore wing stands a little way back from the leading edge, giving rise to a precostal area. The abdomen comprises ten visible terga, of which the tenth bears a pair of unsegmented cerci. The copulatory organ of the male is symmetrical. The female carries an ovipositor by means of which the eggs can be deposited in the soil, between plants, or within them.

A characteristic feature of the Saltatoria is that they are usually capable of producing sounds, generally by rubbing portions of the skeleton against each other or by banging against the substrate with their legs. Special sound-producing organs are frequently found. The Ensifera (bush-crickets and crickets) rub their fore wings together, and most North European species of the Acridoidea (grasshoppers) rub the inner edges of the hind femora against the fore wings at rest. All these various structures are called stridulatory organs, or organs of stridulation.

The Saltatoria frequently bear tympanal (or auditory) organs of some complexity, outwardly visible in the form of a tympanum, more or less covered by an operculum, and thus forming a tympanic cavity. These tympanal organs may be sited in various places: in the Ensifera there is frequently a pair on each side of the fore tibia; in the Acridoidea, on each side of the first segment of the abdomen.

The nymphs closely resemble the adult. The number of nymphal instars may vary widely, not only from one group to another but also within the same species.

Development of the wings is of particular interest. The first buds are normally formed during the second or third instar stages in the form of two small, downward facing lobes from the posterior border of the meso- and metanota. In subsequent instars these buds enlarge, until a remarkable change takes place during one of the final instars. Both wing buds turn upward on their axes at a moult, the medial sides facing outward and the anterior edge upward. The hind wing buds thus cover the fore wing buds. Superficially, a nymph from one of the final instars may resemble an adult with reduced wings, but it is always possible to distinguish a nymph from an adult with reduced wings on the basis of the characteristic appearance and siting of the hind wing buds: fig. 1 compared with e.g. fig. 8. At the final moult, the wings rotate in their points of attachment, finally lying normally along the sides with the posterior edges opposed along the back.

Distribution

There are 45 native species in Northern Europe — a small fraction of the numbers found in Central, Eastern and Southern Europe. Some 150 species are known from

11

Central Europe (Harz, 1957), and some 200 in France (Chopard, 1951). Twenty-eight species occur in the British Isles (Ragge, 1965). The large numbers of species found in Southern and Eastern Europe may be explained on the basis of the higher temperatures and greater number of hours of sunshine in summer.

A total of about 450 species of Ensifera is known in Europe (Harz, 1969), and of Caelifera about 250 species (Harz, 1975).

Collecting, Preparation and Keeping

Collecting

Most North European bush-crickets and grasshoppers are quite easy to catch, the usual method being by net. Where vegetation is dense it may be better to use one's hands if the net will not go in. Another way of collecting is carefully to slide a specimen tube over the animal. This works well with the smaller grasshoppers and groundhoppers. For really satisfactory collecting, the weather needs to be sunny and preferably hot. In cold, wet weather, these insects burrow down into the vegetation and are difficult to find because they are so well camouflaged.

Bush-crickets and grasshoppers can be found virtually anywhere: dunes, moorlands, meadows, ditches, the edges of woods, clearings, overgrown cliffs, and above the tree-line in the mountains of the North — but never in shady woods. Grasshoppers can often be collected in large numbers by sweeping grass and other low vegetation. A lot of the smaller grasshoppers (*Omocestus, Chorthippus*) look very similar among plants, and one may need to have the animal in one's hand in order to identify it. One excellent method is to fill the net with specimens and then to lift them out one by one for identification.

An expert in identifying grasshoppers by their "song" (stridulation) can frequently tell which species inhabit a particular locality without even seeing them. It takes some practice to be able to identify grasshoppers by their "song ", but if one has heard e.g. the Common Field Grasshopper "sing" for some time it is quite easy to determine whether a "song" comes from that particular species or not. Many bush-crickets can also be netted, but in nothing like the sort of numbers typical of the grasshoppers. One usually gets the species living on the ground, plus nymphal instars of forms inhabiting trees and bushes, whose nymphs frequently live on the ground. A very good way of collecting bush-crickets is to knock them down.

There are wide differences between the "songs" of bush-crickets and grasshoppers. The difference is hard to define, but the "song" of the bush-crickets is often louder than that of the grasshoppers, or it may sound like clicking. The best way of finding bush-crickets (e.g. *Tettigonia, Decticus, Pholidoptera*) is to approach them during stridulation. A singing male needs to be approached with the utmost caution, and one should stand quite still if he stops. One can get very close in this manner, although it may still be difficult actually to spot it, since they are so well camouflaged. What often gives them away is wing movement. The next thing to do is to search very carefully to see if there is a female in the vicinity, since it can be quite difficult to get hold of them. If one is lucky, one can go right up to it and take it in one's hands, or tip it into the net. If it is disturbed, it will drop down into denser vegetation and disappear from view. They very seldom fly away.

The singing of bush-crickets and grasshoppers usually starts at about the summer solstice, when bird song is on the decline.

When collecting, one ought if possible to get specimens of both sexes, since this will frequently facilitate identification.

Preparation

Killing may be carried out in the normal way in a potassium cyanide tube or one containing acetic ether or similar. Bush-crickets and grasshoppers usually regurgitate during collecting and killing, so it is best not to put too many specimens into the one killing tube. As with beetles and butterflies, grasshoppers can be dried and kept for subsequent maceration (softening). If collecting on a long expedition, it is best to put one's catch into specimen tubes closed with cotton wool stoppers. The tubes may even be left open for 24 hours or so first, to let the animals dry out better. One animal is put into each tube, and the cotton wool stoppers are moistened with acetic ether, the vapour from which kills off any fungal spores in the tube. The specimens ought to fit the tube quite tightly since, if allowed to rattle about, they can easily lose bits off their legs or antennae. Nymphs are best preserved in alcohol because of their thinner integument. They become highly flexible in a glycerol/alcohol mixture.

Bush-crickets and grasshoppers are generally mounted on a pin through the posterior region of the pronotum, a little to the right of the central ridge or keel. The wings should be spread, since many species are identifiable on the basis of the venation and coloration of the hind wing. As a space-saving measure, one can spread the wings on one side only: usually the right. In the larger species (e.g. *Tettigonia, Decticus*) the soft parts should be removed from the abdominal cavity and replaced by hygroscopic cotton wool. This is most easily done by cutting a slit up the underside of the body, taking care not to damage the subgenital plate. This should be done shortly after killing, otherwise the abdomen may decompose or shrivel. In grasshoppers, the hind legs should be placed so as to ensure visibility of the auditory organ on the first segment of the abdomen. It may sometimes prove necessary to alter the positions of e.g. legs or wings in order to observe some detail necessary for identification. Rapid softening can be achieved by injecting Barber's Fluid into the thorax with a thin needle. The extremities can then be moved in a few minutes.

In the Gomphocerinae it may be necessary for purposes of identification to know the numbers of stridulatory pegs on the inside of the hind femora. These are the pegs which are rubbed against the fore wings during stridulation. To count them, the hind leg can be snipped off and subsequently glued to a small piece of cardboard with the inner edge up. It is a good idea to record the number on the piece of card. A method of preparation making it clearly possible to determine the number of pegs is as follows. The hind femur is soaked in 5% potassium hydroxide solution for about 24 hours, after which it is rinsed in distilled water for a day or so (change the water once or twice). It is then placed in 70% alcohol, after which the denticles can clearly be seen under low power ($\times$ 25) magnification. A permanent preparation can be made by setting the leg in Canada balsam or immersing it in glycerine in a microtube attached to the pin.

In bush-crickets, the titillator is an identificatory feature. It can be observed by bending the subgenital plate (see under Ensifera) downwards, and it is easy to extract with forceps when cut loose from behind with a scalpel or scissors. In dried specimens, the tip of the abdomen can be dipped in 50% alcohol for an hour or so, after which the preparation can be made without damaging the rest of the animal. Another method of softening the abdomen is by dripping boiling water on it — taking care not to soak the whole specimen — and a third is to inject Barber's Fluid. A simular method is used for extracting the epiphallus in grasshoppers: see under Caelifera.

Keeping

It is easy to keep bush-crickets, crickets and grasshoppers alive in cages. They are very lively, and it is fascinating to follow their development from the moment they emerge from the egg as tiny nymphal instar, and through their life.

The bush-crickets are chiefly carnivores, although they do also ingest a certain amount of plant food. The larger species can be fed on meal worms, small grasshoppers etc., whilst smaller species require aphids, small larvae etc. They enjoy tender, juicy pieces of vegetation, such as petals, dandelion leaves or little bits of apple.

The grasshoppers are herbivorous. The thing to do is to put a grassy piece of turf in their cage, and renew it when they have eaten all the grass off it.

The bush-crickets and grasshoppers of Northern Europe die off in the autumn, only their eggs surviving the winter. Putting eggs laid in the soil into the deep freeze will break their diapause. In the case of the Common Field Grasshopper, eggs were kept deep frozen for a month, and the nymphs hatched the following February.

Key to suborders of Saltatoria

1   Antennae longer than body; if shorter *(Gryllotalpa)* the fore legs strongly modified for digging. Tympanal organ, if present, near base of fore tibia. Stridulatory organ, if present, at base of the fore wings (tegmina); present only in males. Ovipositor long and sword-shaped, adapted for boring into earth, rotten wood etc. .... Ensifera (p. 15)

–   Antennae mostly not longer than the head and pronotum together. Auditory organ (where present) on each side of first abdominal segment. Stridulation, if performed, by rubbing the hind femur against the fore wing. Ovipositor short, adapted for digging into soil .... Caelifera (p. 58)

SUBORDER ENSIFERA

Head (Fig. 2)

The structure of the sector between the antennae and the eyes, where the vertex meets the frons, is of special interest. The vertex frequently rises to a greater or lesser prominence, the *fastigium verticis,* between the antennae. This is regarded as having originally had the character of a paired tubercle, as in the Rhaphidophoridae (Fig. 27). Fuses later in development to an unpaired, usually rigid, *fastigium verticis.* This may be narrow and protuberant, and spinous or conical in form, as in e.g. Conocephalinae, Meconeminae and Tettigoniinae, or wide and less prominent, as in the Decticinae. In the Gryllidae it is usually only slightly protuberant and rounded, even though it can be convexly domed. The upper part of the frons carries a more or less prominent *fastigium frontis,* meeting the *fastigium verticis* between the antennae. The ocellus is in the middle of it, and it may terminate dorsally in a small tubercle, resulting in each of the two fastigia bearing a tubercle where they meet *(Leptophyes).* The two fastigia may be completely fused, as in the Grylloidea. The ocellus is absent in *Gryllotalpa.*

The compound eyes may be prominent and spherical, as in the Phaneropterinae and the Tettigoniinae; adjacent and oval as in the Decticinae, or adjacent and circular as in *Gryllus.* A feature characteristic of most Ensifera is the long, filiform antennae, extending backwards beyond the apex of the abdomen and consisting of numerous segments. Antennae twice or even thrice the length of the body are by no means uncommon. Such very long antennae can be seen in the Rhaphidophoridae (Fig. 28),

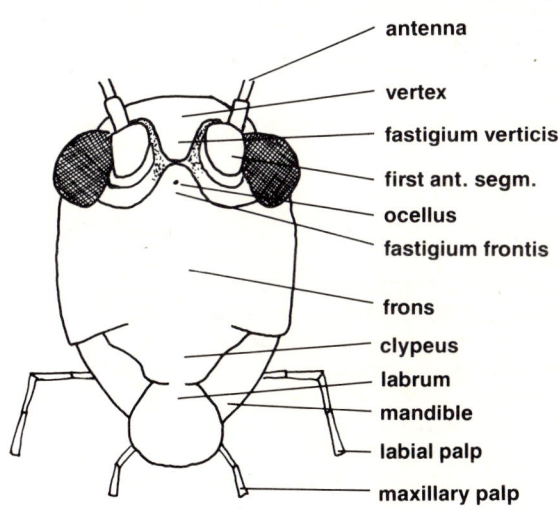

Fig. 2. Head of a bush-cricket (*Tettigonia* sp.).

where they may be an adaptation to these species' habitat in the dark (cave forms: nocturnal in greenhouses). An example of a species with very short antennae is *Gryllotalpa,* an adaptation to its subterranean habit. The antennae are usually inserted between the eyes near their inferior margin, and may be closer to or further from each other, depending on the width of the *fastigia verticis* and *frontis.*

The biting mouth-parts comprise the labrum, mandibles, maxillae — each maxillus with a five-segmented palp — and the labium, with a pair of three-segmented palps.

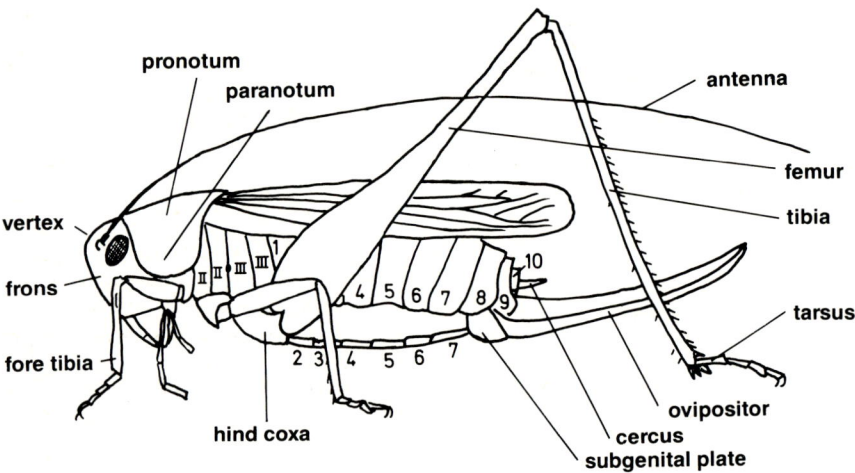

Fig. 3. External morphology of a bush-cricket (*Decticus* sp.) ♀. — II = mesothorax; III = metathorax; 2-10 = abdominal segments.

Thorax, legs and wing (Figs 3-7)

The characteristic saddle-shape of the pronotum in Saltatoria has already been mentioned. It is most commonly somewhat extended posteriorly, over the mesothorax, frequently concealing the insertion of the fore wings. There may be a central keel on the pronotum, but it is not usually particularly prominent. The paranota may be more or less vertical, often terminating in a rim-like edge.

The prosternum is quite broad, with widely separated coxae (except in *Gryllotalpa),* and there may be a pair of downward-pointing spines between them *(Tettigonia).* More or less the same applies to the meso- and metasterna, where the spines have evolved to plate-like lobes, one pair on each segment.

The legs comprise a coxa, trochanter, femur, tibia and tarsus. The first and second pairs of legs are of about the same size and are simple except in specialised species *(Gryllotalpa),* whilst the hind legs are strong and adapted for jumping.

The fore tibia carries the tympanal organ, with a tympanum on the side of each fore tibia. In certain subfamilies (Meconeminae, Phaneropterinae), the tympanum is very obvious, but in most other subfamilies it is partly covered by a flap, the entry to the tympanum being visible as two longitudinal slits on the upper surface of the protibia.

The wings are well developed in most Ensifera, though there are species of the subfamilies Phaneropterinae and Decticinae with shorter or highly reduced wings. In species with much reduced wings, there are frequently small vestiges to be seen in females, whilst the wings of the male are not so reduced as to have lost the stridulatory organ, as seen in e.g. *Pholidoptera griseoaptera*. Most species belonging to the Ensifera are rarely seen in flight, even species in which the wings are so long that they extend backwards beyond the apex of the abdomen. The fore wings (tegmina; called elytra by Harz) in Scandinavian species are generally long, narrow and leathery (Tettigonioidea) or short, wide and leathery (Grylloidea). The hind wings (alea) are folded under the fore wings at rest. They are transparent and thin, and are divided into two sections: — the narrow, pre-anal section and the extensive anal section, which is folded in pleats under the pre-anal section at rest.

At rest, the fore wings are angled slightly posterior to Cu_1, resulting in the wing consisting of a lateral portion, the pre-cubital section, and a dorsal portion, the post-cubital section. The two sections are practically at right angles, as a result of which only the lateral, pre-cubital portion can be seen in Fig. 3. The dorsally placed post-cubital portions of the two fore wings overlap behind the pronotum, reciprocal movements of the tegmina against each other bringing about stridulation. The stridulatory organ is only found in the male sex in Northern European species. The wings are raised slightly during stridulation. The veins involved in the stridulatory organ are barely discernible. One moot point has been whether the modifications which have

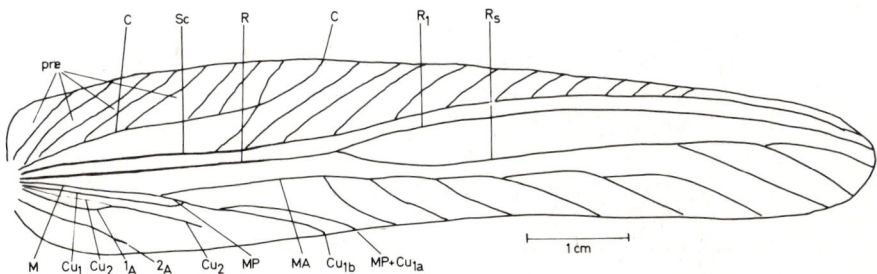

Fig. 4. Left fore wing of ♀ of *Tettigonia viridissima* (L.). — C = costa; Sc = subcosta; R_1 = first radial vein; R_s = radial sector; M = medial vein; MA = anterior medial vein; MP = posterior medial vein; Cu_1 = first cubital vein; Cu_2 = second cubital vein; 1A = first anal vein; 2A = second anal vein. The fore wing has a leaf-like appearance, because C and Sc are sited close together along the middle of the wing, and also because the large precostal area (pre) is provided with a number of additional veins arising from the costa.

taken place in Tettigonioidea are the same as those seen in the Grylloidea. Ragge (1965) will be followed in the present publication.

The stridulatory organ (Fig. 5) consists of two parts: — a pars stridens (stridulatory rib) on Cu_2 (others claim that this should be interpreted as 1A), carrying a series of parallel cross-ribs, and a plectrum on the posterior edge. The noise is brought about by the plectrum running across the pars stridens when the wings are rubbed together, the sound propagating to the wings themselves, where it is amplified.

In the Tettigonioidea the post-cubital portion is small, and the pars stridens is on the left fore wing with the parallel ribs downwards, whilst the plectrum is on the right fore wing. In addition, there is a large, practically circular, very clear, transparent membrane, the speculum (mirror) on the right fore wing. A corresponding structure

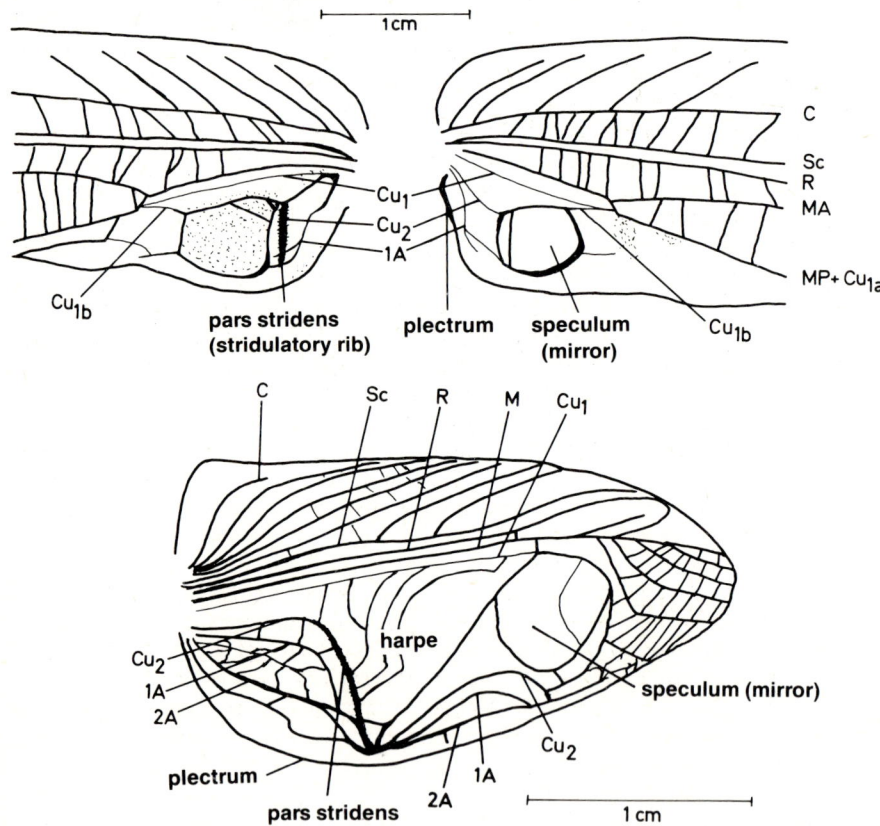

Fig. 5. Above: base of fore wings of ♂ of *Tettigonia viridissima* (L.). — Below: right fore wing of ♂ of *Acheta domestica* (L.). Abbreviations as in Fig. 4.

is present on the left fore wing, but it is pigmented and not transparent like the speculum. In the Tettigonioidea the post-cubital portion of the left fore wing lies over that of its right counterpart.

In the Grylloidea, the post-cubital portion is large, making the wings wider, in addition to which the fore wings are shorter than in the Tettigonioidea; as a result of this, crickets resemble small boxes. In the Grylloidea, both fore wings have a pars stridens and a plectrum, but — unlike in Tettigonioidea — the right fore wing practically always lies above its left counterpart. In addition, the stridulatory organ has moved outwards, almost to the middle of the wing.

Abdomen

This is practically always quite large and broadly attached to the thorax. It comprises ten visible terga. The tergum of segment 10 bears a pair of unsegmented cerci, which are short and inflexible in the Tettigonioidea, but long and flexible in the Gryllacridoidea and Grylloidea. The male has nine sterna, of which sternum 9 is the subgenital plate, bearing two styli in the Tettigonioidea (although not in the Phanopterinae) and Gryllacridoidea. The styli are absent in the Grylloidea. The female has seven sterna, in addition to which there is frequently a small subgenital plate at the base of the ovipositor.

The anus is surrounded by three valves; an upper epiproct and two lower paraprocts, which may extend a greater or lesser distance under tergum 10.

The copulatory organ has no penis as such but consists of symmetrical membranous valves surrounding the opening to the ejaculatory duct on abdominal segment 9. Amongst the copulatory organs we must also include the frequently occurring sclerotised parts (pseudoepiphallus; epiphallus *alias* pseudosternite) attached behind and above the membranous valves. In the Tettigonioidea the pseudoepiphallus is called the titillator and consists of two free or only partially fused rods attached at the back and extending out over the valves. The copulatory organ is normally hidden behind the subgenital plate but it can be observed by bending it down. The titillator is of great significance in identification.

The ovipositor originates on segments 8 and 9. It consists of three pairs of valves in the Tettigonioidea and Gryllacridoidea: a superior pair, an inferior pair and an inner pair. The upper and lower valves are on the outside and are long, concealing the much shorter inner valves. The ovipositor in the Gryllidae only consists of the outer valves, the inner pair being absent. It can be of greater or lesser length, broad or narrow, highly curved or practically straight. The shape is frequently of great significance in systematics. In the Gryllotalpidae the ovipositor is completely reduced.

Development

The number of nymphal instars may vary, not just from one species to another but apparently within the same species. The number of nymphal instars is usually five or six.

The so-called "vermiform nymph" appears at hatching; it is surrounded by a thin, transparent membrane which erupts when the nymph arrives at the surface of the soil. It carries out worm-like motions to reach the surface. It has been agreed that the nymphal stage which arises after eruption of the skin of the vermiform nymph should be called the first instar (Nymph 1). The first wing primordia appear in Nymph 2 or 3, the wing buds becoming rotated upwards in the two last instars. Nymphal development: see Richards (1958).

The eggs are laid singly and vary in form. They are usually oval and more or less cylindrical. In the Phaneropterinae the eggs are very flat. For structure and placement of eggs, see: Hartley (1964). Descriptions of eggs and nymphs are found in Ragge (1965).

The sizes quoted for imagines were measured on Scandinavian specimens, and in females the length of the ovipositor is not included in the overall length.

Times of hatching of nymphs and of development of imagines and dates of final observations of imagines in the autumn were determined in Scandinavian specimens.

Habitat preferences

Species of Ensifera are usually to be met with in trees, bushes, shrubs and tall grass. *Tettigonia viridissima* and *Meconema thalassinum* belong to the species which reach highest up trees. *T. viridissima* may, however, be found closer to ground level in shrubs in which *Leptophyes punctatissima, T. cantans, Decticus verrucivorus* and *Pholidoptera griseoaptera* are also found. On lighter soils, carrying grass, heather and scrub, *Decticus verrucivorus, Metrioptera brachyptera, M. roeseli, M. bicolor, Platycleis albopunctata* and *Gryllus campestris* may be found. The following species may occur in dunes: — *Platycleis albopunctata, Decticus verrucivorus* and *Conocephalus dorsalis,* although this latter is commoner in littoral meadows. *Metrioptera brachyptera* and *M. roeseli* may also be found on wetter soils, e.g. in bogs. The highly specialised *Gryllotalpa gryllotalpa,* the mole cricket, is the mole amongst the Saltatoria.

Diet

The diet is primarily animal in origin and — depending on size — bush-crickets may consume grasshoppers, larvae, aphids etc. The fore legs are frequently used both for catching the prey and subsequently holding it while it is being eaten. The mandibles may also be used to grasp smaller animals with. If a bush-cricket is disturbed while eating, it will carry its prey away with it in its mandibles. It was once believed that the Ensifera were exclusively carnivorous, but is has since transpired that they eat plant food from time to time. This particularly applies to juicy plant parts (petals, dandelion leaves, etc.). One species which is largely vegetarian is *Leptophyes punctatissima.*

Stridulation

The sound produced by stridulation varies widely, so it is impossible to give one, over-all description of it. The noise may be very loud and confluent, or it may sound like ticking, clicking or faint rustling or crackling. In Scandinavian species, only males sing. Their song, the call, is the only sound produced by bush-crickets. It has a frequency of between 4,000 and 100,000 c/s (cycles per second, or Hz (Hertz)). The human ear is normally sensitive to sounds between 20 and 20,000 c/s, i.e. we can normally only hear the lower portion of their range of frequencies, which we experience as very high notes.

The stridulation of crickets (Grylloidea) frequently sounds like a faint whistling or flute-like sound. Their frequencies are in the 2,000-6,000 c/s range, i.e. much lower than in bush-crickets, and all lying within the range to which the human ear is sensitive. Crickets can sing in various manners. The male first sings his call, attracting the female to him.

Stridulation can be described in various ways, the commonest being on the basis of the number of vibrations, ticks etc. per second, and whether there is a falling or a rising note. These descriptions are also used here. Oscillograms of the various species' stridulation have also been recorded in order to give a visual impression of the sound. The terminology applied by Elsner (1974) is used here under the various descriptions: see under Caelifera.

Key to superfamilies of Ensifera

1   Tarsi with three segments. Body cylindrical. Postcubital portion of fore wings large . . . . Grylloidea (p. 50)
–   Tarsi with four segments. Body laterally compressed 2
2 (1) Tarsi long and compressed. Wings absent. Tympanal organs absent . . . . Gryllacridoidea (p. 47)
–   Tarsi short and flattened. Wings present, though sometimes rudimentary. Tympanal organs usually present . . . Tettigonioidea (p. 21)

Superfamily Tettigonioidea

(Bush-crickets)

Antennae long and thin. Tarsus flattened, 4-segmented. Fore tibia usually bearing tympanal organs. Wings well developed, or more or less reduced, but never completely absent. Postcubital portion (= dorsal portion) of fore wings small, bearing the stridulatory organ. Left fore wing carrying the pars stridens and overlapping the right fore wing, which carries the plectrum and speculum. Cerci rather short and inflexible. Styli normally present on subgenital plate in male. Copulatory organ symmetrical, consisting of membranous valves; titillator frequently present. Ovipositor generally large, consisting of three pairs of valves.

Only one family.

21

Family Tettigoniidae

Key to subfamilies of Tettigoniidae

1   First and second tarsal segments grooved laterally (Figs 6B and 6C). Fore and mid tibiae without a longitudinal groove. ♂ subgenital plate with a pair of styli.............................. 2

–   First and second tarsal segments laterally without a groove (Fig. 6A). Fore and mid tibiae with a dorsal longitudinal groove. ♂ subgenital plate without styli .... Phaneropterinae (p. 23)

2 (1) Basal segment of hind tarsus without movable flaps.................... 3

–   Basal segment of hind tarsus with two movable flaps (Fig. 6C) .... Decticinae (p. 35)

3 (2) Fore tibia at apex with a dorsal spine on the outside (Fig. 7C & D)..................................... Tettigoninae (p. 31)

–   Fore tibia at apex without a dorsal spine on the outside 4

4 (3) Tympanum with opening quite unrestricted (Fig. 7A) ... Meconeminae (p. 26)

–   Auditory organ with opening reduced to a slit (Fig. 7B) Conocephalinae (p. 28)

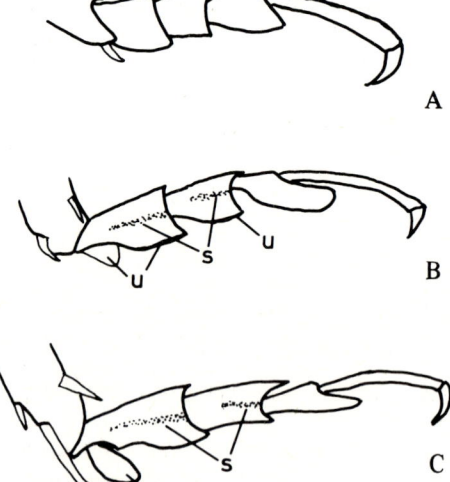

Fig. 6. Hind tarsus of A: *Leptophyes punctatissima* (Bosc); B: *Tettigonia viridissima* (L.); C: *Decticus verrucivorus* (L.). — s = lateral groove; u = unmovable flap; h = movable flap.

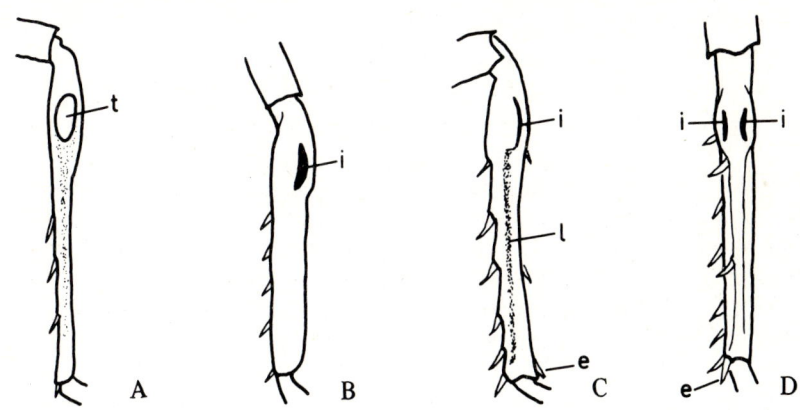

Fig. 7. Fore tibia of A: *Meconema thalissinum* (De Geer); B: *Conocephalus dorsalis* (Latr.); C & D: *Tettigonia viridissima* (L.). — A-C in lateral view, D in dorsal view. — e = dorsal spine; i = entrance to auditory organ; l = longitudinal groove; t = tympanum.

SUBFAMILY PHANEROPTERINAE

Head rounded, eyes small and protuberant. Fore and mid tibiae with dorsal longitudinal groove. Fore tibia with unrestricted tympanum. Tarsal segments 1 & 2 without lateral grooves. No styli in male. European species generally long-legged animals with reduced wings.

Only one genus in N. Europe. Group widely represented in Central, S. & E. Europe.

Genus *Leptophyes* Fieber, 1853

Leptophyes Fieber, 1853, Lotos 3: 174.
Type-species: *Barbitistes albovittata* Kolenati, 1833.

Fastigia verticis and *frontis* each bearing a bud-like swelling where they meet between the eyes. Fore tibia mostly twice as long as pronotum. Fore wings reduced to small flaps. Cerci of male straight, bent like a hook towards the tip. Ovipositor short, broad at base, then suddenly compressed and bent sharply upwards like a scimitar; finely denticulate apically.

1. **Leptophyes punctatissima** (Bosc, 1792)
   Figs 6A, 8, 9.

Locusta punctatissima Bosc, 1792, Act. Soc. Hist. Nat. Paris 1: 44.
Eng.: Speckled Bush-cricket; Dan.: Krumknivgræshoppen; Fin.: Tarhahepokatti; Sw.: Lövvårtbitaren.

Hind border of pronotum bent slightly upwards, extending a little way over the fore wings. These are slightly larger in the male than in the female. Hind wings smaller than fore wings in both sexes. Green, with fine black-red speckling. Dorsal surface of abdomen brownish in colour. Antennae yellowish green, with black rings. A highly attractive animal in life; frequently loses bright coloration on preparation, although the red-black speckling may often still be seen. Length: ♂ 9-16 mm, ♀ 11-17 mm, ovipositor 6.5-7.5 mm.

Distribution. Occurs sporadically in Denmark and southern Fennoscandia. In Denmark, found especially in the south of the country: on Møn, Lolland, Falster, Funen, along Flensborg Fjord (SJ) and in the Viborg region (EJ), towards Aalborg (NEJ); also near Copenhagen and on Bornholm. — In Sweden and Norway it is found in the coastal areas bordering the Kattegat and Skagerrak, from Kullen in the south to Vest-Agder in the northwest. Also occurs on Öland and Gotland. — In Schleswig-Holstein, has been found near Eckernförde and Heide. — From Ireland and Great Britain, France, the Netherlands and Belgium eastwards to the European part of the USSR; in Southern Europe from Spain to Greece and Yugoslavia.

Biology. Inhabits bushes, herbaceous plants and the lower portions of trees. Found in the Scandinavian countries on blackberry bushes and rows of hops, in oak copses,

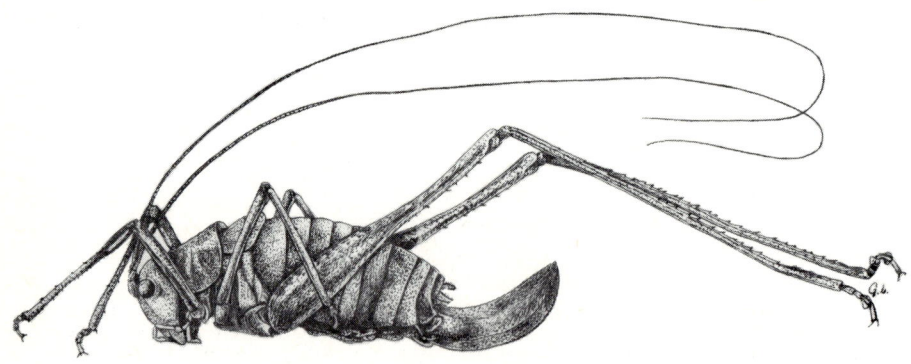

Fig. 8. *Leptophyes punctatissima* (Bosc) ♀.
Length 11-17 mm.

24

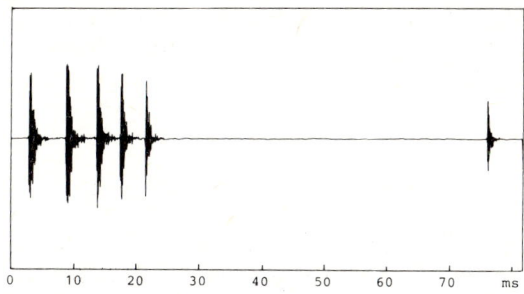

Fig. 9. Oscillogram of the song of *Leptophyes punctatissima* (Bosc). — S, Sk.: Förslöv, October 1979. (From Ahlén & Degn, 1980).

on oaktrees and on Arbor vitae *(Thuja)* hedges. In late summer, frequently found several metres up trees, perhaps to lay eggs. Appears primarily to be a dusk and evening animal (Harz, 1957), although it has been established that it may stridulate both night and day in August and well into October (Ahlén & Degn, 1980). Herbivore, but probably takes small animals as well. Has been seen on plants, e.g. roses, which bear the marks of its bite.

The eggs are orange to brown, with a whitish surface layer; they are oval and very flat, 3.5 mm in length and 1.5 mm in width, laid singly in stems of plants and cracks in bark of young trees and bushes. Eggs have a good chance of being distributed with cut-off branches and herbaceous plants from nurseries. During laying, the female doubles up completely, grasping the ovipositor near its base with her mouth-parts (Harz, 1957).

The nymph lives in the base of vegetation and is considerably more active by day than the imago. The first nymphs hatch in May, and nymphs have been observed right through to the end of August. Six nymphal instars have been observed. The imago is found from the middle of July until the beginning of October.

Stridulation. The song has been described many times, but it is a moot point whether it is audible to the human ear. In Central European specimens, stridulation has been described as a series of individual tones which may be repeated with various rhythms depending on temperature. Stridulation is quite faint, and can be heard as a weak rustling or crackling at some 75 cms. distance (Fabre, 1953; Harz, 1957). Sandhall & Ander (1978) described the song as a short simple "sr"-sound. Ahlén & Degn (1980) analysed stridulation emanating from crickets out of doors in Jutland, Funen, Skåne and Gotland. They were unable to hear the song, and they give the following description. "The loudest portion of the song is usually about 40 Kc/s, varying according to our recordings between 33 and 43 Kc/s from individual to individual. The fact that the sound is not as faint as was previously supposed can be demonstrated with the aid of an ultrasonic detector (QNCS 100), where the chirping can be heard at a distance of over 25 metres." Individual chirps consist of from five to seven bursts of sound, fre-

25

quently at 3-7 millisecond intervals, usually terminating with a peak some 50 ms later. The entire chirps thus lasts some 75 ms, including the terminal peak (Ahlén & Degn, 1980).

SUBFAMILY MECONEMINAE

Eyes small, spherical and protuberant. A furrow or groove on each side of the fore tibia, terminating above in an oval, unrestricted tympanum. Distal end of fore tibia lacks dorsal spines. Stridulatory organ generally rudimentary.

Four species known in Europe, belonging to three genera. Only one genus in our area.

Genus *Meconema* Serville, 1831

Meconema Serville, 1831, Ann. Sci. Nat. 22: 157.
   Type-species: *Locusta thalassina* De Geer, 1773.

Antennae approximately twice as long as body. Fastigium verticis conically swollen between the eyes. Pronotum cylindrical, without keel. Legs long and thin. Wings well developed or reduced. Cerci long, straight, brush-like and slightly curved in male. Ovipositor approximately same length as abdomen, non-denticulate and slightly curved.

Two species in Europe.

2. *Meconema thalassinum* (De Geer, 1773)
   Figs. 7A, 10.

Locusta thalassina De Geer, 1773, Mém. Ins. 3: 433.
Eng.: Oak Bush-cricket; Da.: Egegræshoppen; Sw.: Ekvårtbitaren.

Well-developed wings, extending to end of hind femora. Delicate shade of light green; rarely yellowish. Antennae yellow with brown rings and spots. A yellow to brown stripe extends from the vertex over the pronotum to the posterior margin of the fore wings. Two black spots on rearmost portion of pronotum. Length: ♂ 12-15 mm; ♀ 11-16 mm; ovipositor 8-9 mm.

Distribution. Widely distributed and common in the southern and eastern regions of Denmark, and in Skåne in Sweden. More scattered in the rest of S. Sweden. — Occasionally reported in Norway: a few records from VE and VAy in the extreme south of the country. — From Ireland and Great Britain to Central Russia, the Crimea and the Caucasus, south to Portugal and northern Spain, the whole of France, Italy, and Yugoslavia.

Biology. A typical arboreal form, most usually occurring on oak but also found on other broad-leafed trees (hazel, lime, apple, elm, sycamore) and even on pine. It appears primarily to be associated with clumps of trees or individual trees in e.g. parks, gardens, avenues and hedges rather than woods and forest margins. May act sluggishly by day. It will attempt to avoid capture by crawling away. Active in the evening and at night, running and jumping about the branches and leaves. Flies readily, frequently seeking light, entering open windows late in summer. Regarded as being more carnivorous than herbivorous, living on small animals (larvae, aphids etc.), fresh leaves and sweet things (cherries, honeydew).

Eggs light brown, oval, somewhat flattened, 3 × 1 mm. Laid in cracks in bark, empty galls etc. This may explain why it prefers oak to other trees and older trees to younger ones.

Nymphs found especially on bushes, tall herbaceous plants or the lower portions of the crowns of trees. The first nymphs hatch in May, and nymphs have been observed right up to August. Five larval instars have been observed. The imago can be found from the middle of August and well on into October.

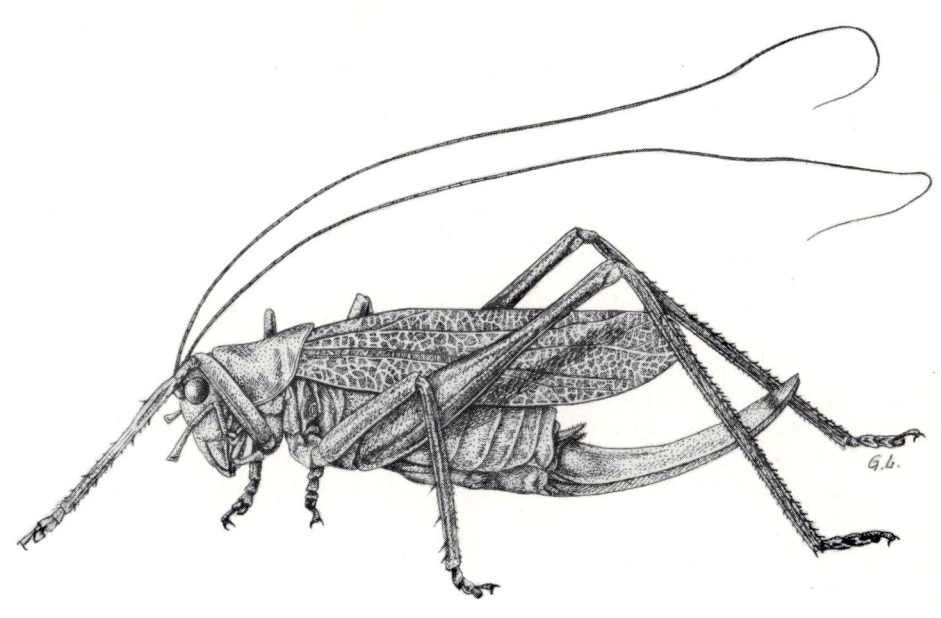

Fig. 10. *Meconema thalassinum* (De Geer) ♀.
Length 11-16 mm.

Stridulation. The male has no stridulatory organ. It produces sounds by drumming its legs on whatever it is standing on, the sound depending on its quality. The usual "song" sounds like "trrrr... trrrr... trrrr... trttt... trrrr...", occasionally audible up to several metres away. While drumming, the head and thorax are bent slightly forward, the wings are lifted straight up at right angles to the body, the hind limbs are pressed against the abdomen, and the abdomen and hind limbs are moved up and down together. Drumming begins in the evening, and in favourable conditions may persist until sunrise. On the approach of a female, the male presses his abdomen flat down and lifts his wings forward, above his head, at an angle of 120° to the abdomen. Copulation can now begin. It lasts 10-15 min. The male lies under the female, holding the tip of the ovipositor fast in his mandibles (Harz, 1957).

SUBFAMILY CONOCEPHALINAE

Eyes small, spherical, protuberant. *Fastigium verticis* projecting, forming a cone or peg with the oblique frons. Tympanun with opening reduced to a slit. Fore tibia round or grooved on one side, without dorsal apical spines. First and second segments of the tarsi with a fine lateral groove on each side.

Only one genus and one species occurs in N. Europe. Two genera (ten species) known to occur in Europe.

Genus *Conocephalus* Thunberg, 1815

Conocephalus Thunberg, 1815, Mém. Acad. St. Petersbg. 5: 214.
   Type-species: *C. bituberculatus* Redtenbacher, 1891.
Xiphidium Serville, 1831; Ann. Sci. Nat. 22.
   Type-species: *X. discolor* (Thunberg, 1815).

Wings well developed (macropterous) or reduced (sub-brachypterous). Hind legs long and thin; hind femora slender in about apical third. Male cerci cone-shaped, straight, each with an inner tooth. Ovipositor very finely crenellated or smooth; its length and curvature vary from species to species.

Only one species in Northern Europe.

3. *Conocephalus dorsalis* (Latreille, 1804)
   Figs 7B, 11, 12.
Locusta dorsalis Latreille, 1804, Hist. Nat. Crust. Ins. 12: 133.

Eng.: Short-winged Cone-head; Da.: Sivgræshoppen; Fin.: Kaislahepokatti; Sw.: Sävvårtbitare.

Fore wings extend to the middle of the abdomen in females and a bit beyond the middle in males. Light green, with a broad, brown to black (or even reddish) dorsal stripe, covering the frons, the dorsal surface of the pronotum and the rearward portion of the fore wings. Eyes brown. Antennae brown with black rings. Fine black spots on legs. Ventral surface of abdomen yellow, green to brown. Ovipositor long, curving smoothly upward. Length: ♂ 11-15 mm; ♀ 12-18 mm; ovipositor 8-9 mm.

Distribution. Found especially near coasts, where the species can be very common. It has been demonstrated on the island of Zealand that it does not follow the vegetation back up a stream (Halleby å), even though the vegetation is the same as on the coast (Holst, 1965). — By the West coast tidal mudflats (Waddenzee) of both Denmark and Schleswig-Holstein. — In Denmark, found near Tipperne (WJ), and round the coasts of E. Jutland, Funen, Zealand, Lolland, Falster, Møn and Bornholm, as well as on the Baltic coast of Schleswig-Holstein and by the R. Elbe. — Sweden: common round the coast of Skåne and northward up the W. coast to Gothenburg. Probably also occurs in Bohuslän, since it is found again in Østfold and Vestfold in Norway. Round the E. coast of Sweden to Uppland; and on Öland and Gotland. — Finland: Alandia, and along the S. coast to the Leningrad area. — England and Wales, France, C. Europe, Yugoslavia and USSR.

Biology. The species is especially associated with marshy vegetation near coasts in littoral meadows, by canals, estuaries and lagoons, and in saltings, but it also occurs in short grass and on both "grey" and "white" dunes. Well camouflaged when perched on stems or blades of grass. Sits pressed close to the plant with its antennae

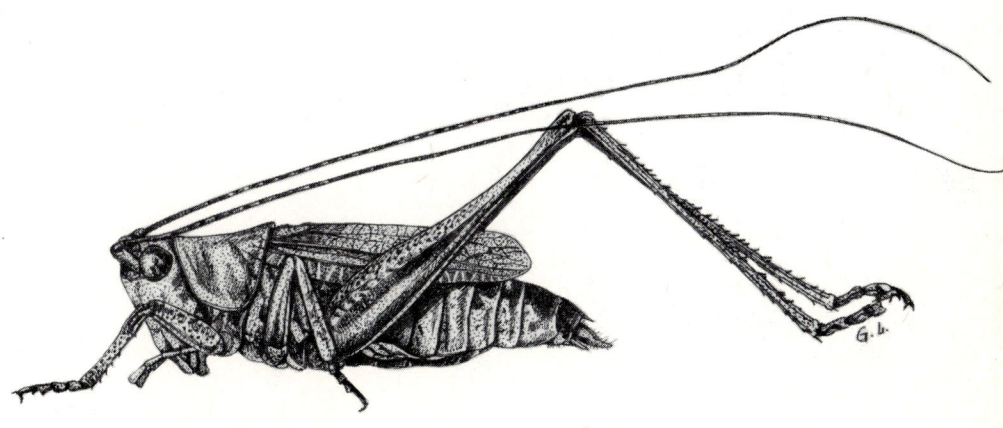

Fig. 11. *Conocephalus dorsalis* (Latr.) ♂.
Length 11-15 mm.

and fore legs stretched forward and the hind pair of legs backward. The green coloration, combined with the brown stripe down the back, make it difficult to see. When disturbed, it shuffles round to the opposite side of the stem or blade of grass. If in acute danger, it jumps to safety. Very lively on hot days, the male usually singing (stridulating) all the time. Lives on small insects (cicadas, flies, midges, larvae) and sappy parts of plants.

The eggs are brownish, elongated and sausage-shaped, 5-6 × 1 mm, laid in the stems of *Juncus, Carex* and possibly *Phragmites*. They can survive for several months in salt water, and it is in this way that the species is thought to have spread from the Continent to the South of England (Warne & Hartley, 1975). Spread along the coasts of Scandinavia may also have taken place by sea, since it can be assumed to have migrated here after the Boreal period, at which time Southern Scandinavia was part of the European mainland.

The first nymphs hatch in May, and nymphs can be found right up to the end of August. This may be due to the fact that the length of the diapause is more temperature-dependent than in other bush-crickets, so the eggs laid latest in the previous year hatch late on (Warne & Hartley, 1975). The imago occurs from July onwards and into October.

Stridulation. Delicate, clear, alternating between two melodies which are repeated at a few seconds' interval without a pause. The one sounds like a whistle, the chirp being long, comprising several syllables, and the other like tickling, the chirp being short, consisting of few syllables.

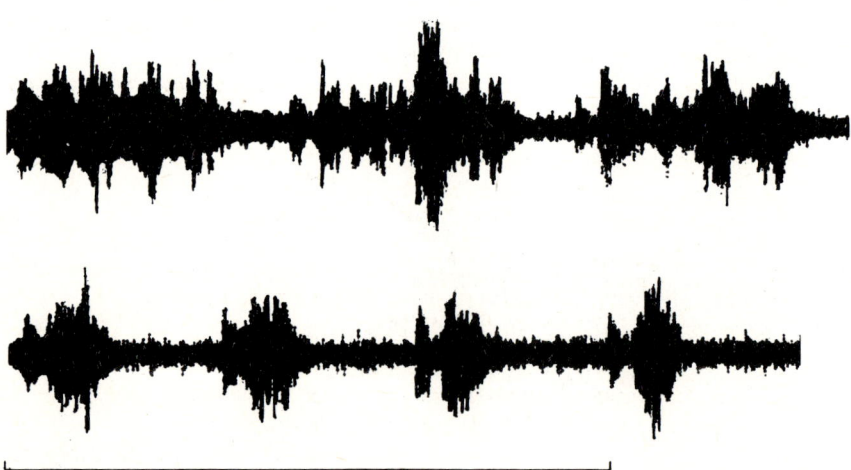

Fig. 12. Oscillogram of the song of *Conocephalus dorsalis* (Latr.). — DK, NWZ: Bjerge Strand, August 1964. Scale: 0.25 sec.

SUBFAMILY TETTIGONIINAE

Eyes small, spherical and protuberant. *Fastigium verticis* very prominent and as wide as, or only slightly wider than, the first segment of the antenna. Tympanum with opening reduced to a slit. Fore tibia grooved on both sides, with three spikes on the dorsal surface, the upper one being sited just below the auditory organ, and the lowest apically. The first and second segments of the tarsi grooved. Hind tarsus with segments 1 & 2 lacking movable flaps but may possess immovable ones.

Only one genus, with two species, in N. Europe. In Europe as a whole, two genera and five species are known.

Genus *Tettigonia* Linnaeus, 1758

Tettigonia Linnaeus, 1758, Syst. Nat. ed. 10: 429.
Type-species: *Gryllus viridissimus* Linnaeus, 1758.

Pronotum smooth, with no lateral edges. Both fore and hind wings longer than the abdomen. Cerci in male straight or slightly curved, both with an inward-pointing bud-like peg. Ovipositor long and either somewhat downward curving or straight.

The genus comprises a few large species.

Key to species of *Tettigonia*

1 Fore wings extend considerably beyond hind femora. Ovipositor curved slightly downwards towards the tip ... 4. *viridissima* (Linnaeus)
– Fore wings only extend a short way beyond the hind femora. Ovipositor straight . . . . 5. *cantans* (Fuessly)

4. ***Tettigonia viridissima*** (Linnaeus, 1758)
Figs 4, 5, 6B, 7C & D, 13.

Gryllus viridissimus Linnaeus, 1758, Syst. Nat. ed. 10: 430.

Eng.: Great Green Bush-cricket; Dan.: Den store grønne løvgræshoppe; Fin.: Lehtohepokatti; Norw.: Grønn lauvgrashoppe; Sw.: Gröna vårtbitaren.

Most of the body green. There is, however, a brown to reddish brown (or even black) stripe down the middle of the back, running from the vertex to the dorsal parts of the fore wings. The fore wings are narrower than in the next species, and the stridulatory organ in the male occupies a sixth of the postcubital portion of the wing. A rest, the fore and hind wings extend well out over the apex of the abdomen. The hind wings are transparent. The antennae are brown or green. Legs sometimes yellowish. Cerci in male long, extending well beyond the styli. Ovipositor curving slightly down towards the tip. Length: ♂ ♀ 25-35 mm; ovipositor 22-25 mm.

Distribution. Denmark: common in the east of Jutland, near the coast, and on the Islands. One single record from south of Hirtshals, NEJ. — In Schleswig-Holstein it has been found in various places, including the west. — In Sweden it has been recorded as being common in various places, extending as far north as Värmland, Västmanland and Uppland. — It has been found in the south of Norway near the coast, extending northwestwards as far as Hordaland. — In Finland it is extensively distributed on Alandia and in the southeasterly regions. — S. England and Wales, Europe, N. Africa, USSR and C. Asia.

Biology. Lives in trees, bushes and the taller herbs (e.g. thistles), where it frequently rests at the top. Rarely seen in flight; prefers to walk. When it feels threatened, will retreat into the vegetation, where it is camouflaged. The author has experienced that it is possible to poke them and irritate them into flight provided the temperature is high enough. The flight commences with a jump. It does not usually fly more than a few metres, but it can fly up to a hundred metres or possibly more. Ander (1947) observed them in flight. The habitat comprises fields, the borders of woodlands, meadows, dunes and gardens. This bush-cricket is most active from the afternoon until quite late at night, at which time stridulation takes place. Mainly carnivorous. The prey is grasped between the legs and the mandibles. Any plant food eaten consists principally of sappy leaves.

The eggs are dark grey, oval, and slightly compressed, 5.3 mm in length and 1.5 mm in width. They are laid in the soil singly or in batches of from two to four. The total number laid is from 70 to 100. The first eggs hatch in May, and nymphs have been observed up to about the end of July. Nymphs remain in lower vegetation than do the adults.

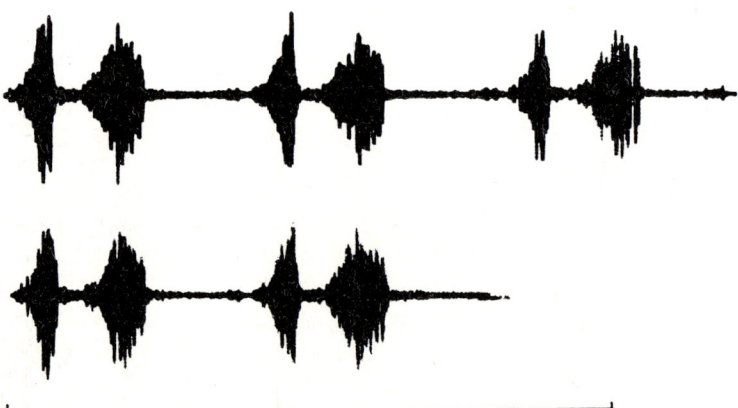

Fig. 13. Oscillogram of the song of *Tettigonia viridissima* (L.). — DK, NEZ: Lundtofte near Lyngby, August 1979. Scale: 0.25 sec.

The number of nymphal instars seems to be variable, between six and nine, eight being generally regarded as the most usual. The imago is found from July to September.

Stridulation. The song is loud, being audible up to a distance of some 200 metres. It consists of a series of loud chirps, which never merge, however high the temperature. Each chirp tends to fall slightly towards the end. The number of chirps is between 12 and 15 per second, the rhythm slowing at lower temperatures. Stridulation can last uninterruptedly for long periods, with intervals lasting from three to twenty seconds. This insect is highly stationary, the same male sitting and singing in the same tree night after night. There are often slight differences between the "voices", and with a bit of practice it is possible to distinguish the individual males by their song. Stridulation starts in the late afternoon, at a time when these insects' activity is increasing, and it may last throughout the night until a few hours before sunrise. It is triggered by the fading daylight and stops when temperatures drop below some 12-15°C (Tetens Nielsen, 1938 & 1970). On exceptionally hot days, stridulation may commence earlier.

Each chirp consists of three syllables: one short, not particularly loud one, most probably produced during wing opening; and two loud ones, most probably produced by spasmodic closure of the wings.

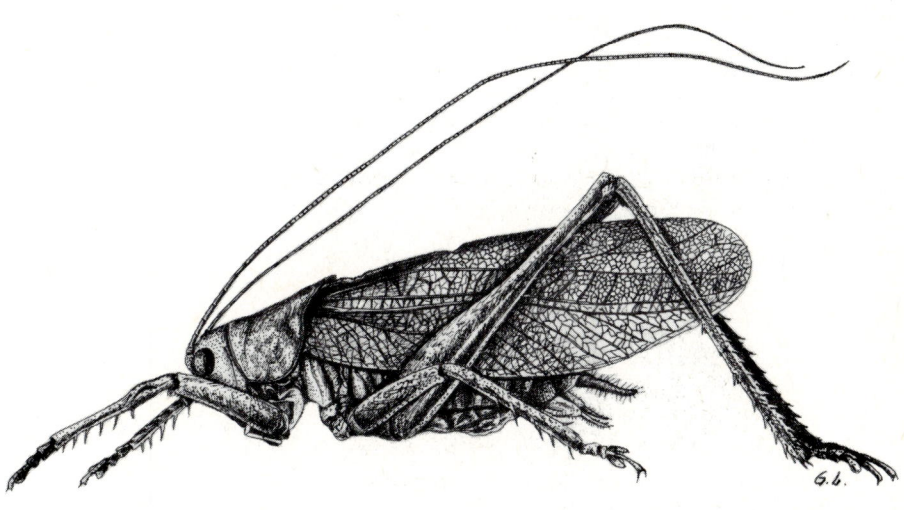

Fig. 14. *Tettigonia cantans* (Fues.) ♂.
Length 22-33 mm.

5. *Tettigonia cantans* (Fuessly, 1775)
   Figs 14, 15.

Gryllus cantans Fuessly, 1775, Verz. Schweiz. Ins.: 23.
Da.: Den syngende græshoppe; Fin.: Idänhepokatti.

Colours resemble those of the previous species: green, with a brown upper surface. Fore wings broad. The stridulatory organ in the male takes up more space in this species than in the previous one, occupying a quarter of the postcubital portions of the fore wings. Fore and hind wings extend only a short way beyond the hind knees. Hind wings transparent. Cerci in male extend only a short way beyond the styli. Ovipositor straight. Length: ♂ ♀ 22-33 mm; ovipositor 22-31 mm.

Distribution. Found in scattered localities in Denmark in eastern South Jutland and East Jutland in the valleys near Vejle and Åbenrå; also on Mols (EJ) and near Herning (WJ). Common in north-western Funen; scattered finds in southern Funen. — Not recorded from Sweden and Norway. — Recorded as common in Holstein at a number of localities north of Lübeck and southward to Hamburg. — In Finland, it has been recorded in the southeastern region of the country, its zone of distribution extending into the neighbouring areas of the Soviet Union. — A Euro-Siberian species, mainly in mountainous areas (the Pyrenees, the Alps, the Apennines, and the mountains of the Balkans).

Biology. Found in Denmark and Holstein in gardens, hedges, and at the edges of woods and ditches. May be found sitting from 1 to 3 metres up in the vegetation, e.g. in bushes and trees and the tops of the taller herbaceous plants such as thistles and potatoes. In Finland, it has been found on grasses and herbaceous plants, which may be mixed with bushes and low trees, practically only in association with agriculture. The southeastern portion of Finland was not colonised by man until the 10th Century, and it was virtually unpopulated until the 16th Century. This species cannot survive in thick forest, and it is likely that it was only on the introduction of agriculture that it migrated in from the East (Panelius, 1978).

Fig. 15. Oscillogram of the song of *Tettigonia cantans* (Fues.). — DK, F: Brenderup, September 1964. Scale: 0.25 sec.

Stridulation. Stridulation starts early on in the afternoon and may last until after midnight, but it will cease if temperatures drop below about 10° C (Panelius, 1978).

The song is very loud and can be heard up to several hundred metres away. At high temperatures, individual chirps tend to merge into one long, piercing, uninterrupted call which can go on for from one to six seconds. At lower temperatures it may be difficult to hear the difference between the song of *T. viridissima* and that of *T. cantans,* but *cantans* has an impure, jangling undertone, without the falling note on each chirp so characteristic of *viridissima.* Each chirp consists of two syllables: a short, not particularly loud one, and a louder one.

SUBFAMILY DECTICINAE

Eyes large, oval and only slightly protuberant. *Fastigium verticis* rounded and much wider than first antennal segment. Fore tibia grooved on both sides. Tympanal openings reduced to slits. First and second segments of the tarsi laterally grooved. First segment of metatarsus has two long movable flaps on the underside. The wings may be well developed, but may also be reduced.

Four genera and six species known in N. Europe. 28 genera and some 160 species known in Europe as a whole.

Key to genera of Decticinae

1   Apex of fore wings (tegmina) extend beyond apex of abdomen 2
–   Apex of fore wings not extending beyound apex of abdomen 3
2 (1) Pronotum with a median keel throughout its entire length. Length 24 mm or more. . . . . *Decticus* Serville (p. 35)
–   Hind portion of pronotum having a median keel. Length seldom over 20 mm. . . . . *Platycleis* Fieber (p. 38)
3 (2) Fore wings covering about half the abdomen . . . *Metrioptera* Wesmael (p. 39)
–   Fore wings short and scale-like . . . . *Pholidoptera* Wesmael (p. 45)

Genus *Decticus* Serville, 1831

Decticus Serville, 1831, Ann.Sci.nat. 22: 155.
    Type-species: *Gryllus verrucivorus* Linnaeus, 1758.

Large insects. Pronotum flat on the upper surface, with a median keel running its entire length. No spines on prosternum. Four ventral spurs on the hind tibiae. Movable flaps of hind tarsi shorter than first tarsal segment. Cerci in male each with an inward-facing peg. Ovipositor curves slightly upwards and crenulated at the apex.

6. *Decticus verrucivorus* (Linnaeus, 1758)
Figs 1, 6C, 16, 17.

Gryllus verrucivorus Linnaeus, 1758, Syst.Nat.ed. 10: 431.
Eng.: Wart-biter; Dan.: Vortebideren; Fin.: Niittyhepokatti; Norw.: Vortebiter; Sw.: Stora vårtbitaren.

Wings extending less than one third of their length out over the apex of the abdomen. In females prior to laying, however, the abdomen can be very swollen, their wings then only extending to about the apex of the abdomen. Lateral keels on pronotum. Very variable in colour: generally green with brown spots on the wings and abdomen, but they can be completely green, or more or less brown. Black ones and even red ones have been reported. Length: ♂ ♀ 24-45 mm; ovipositor 17-26 mm.

Distribution. This species has been found generally throughout Denmark as well as in Schleswig-Holstein in suitable localities. — In Sweden, it extends as far north as Dalarna, and a few finds have been reported in Norrland and Lapland. — In Norway, it has only been found in the most southerly parts, extending to Rogaland in the west. — It has been observed throughout Finland, except for the far north, and it has also been found in the neighbouring areas of the Soviet Union. It seems to have become rare in Denmark and Skåne in recent years. — A widely distributed palaearctic species.

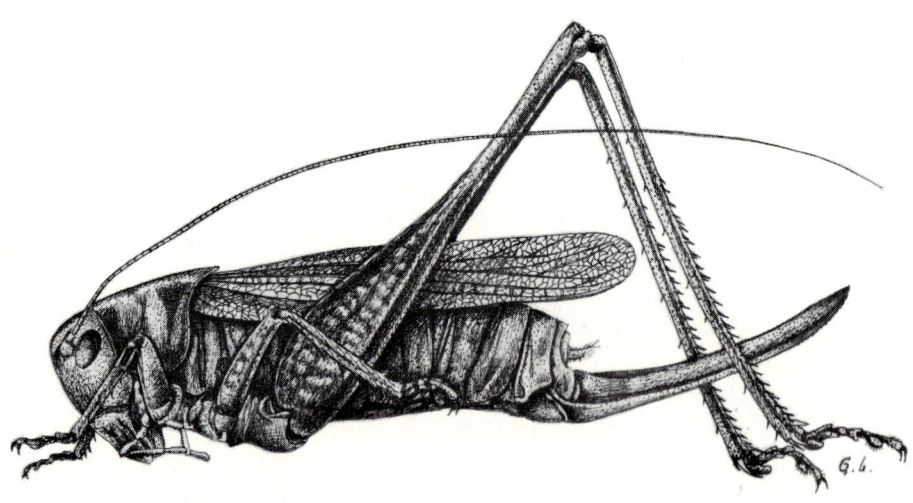

Fig. 16. *Decticus verrucivorus* (L.) ♀.
Length 24-45 mm.

Biology. Occurs in dry and damp places in fields, moorlands, meadows and by road-sides. Particularly fond of low vegetation; does not climb particularly high up trees or bushes. Rarely seen in flight; prefers to crawl away when in danger. Has been observed in flight with its hind legs trailing; presumably flies longer distances at higher temperatures (Ander, 1947).

D. verrucivorus is not especially carnivorous. The name probably originates in Central Sweden, where Linnaeus observed the way in which the peasants allowed these insects to bite holes in their blisters and drink the lymph which seeped out. It may capture larvae, grasshoppers etc., which it grasps with its fore legs or its mandibles. Should it walk off, it carries its prey in its mandibles. Its plant food mainly consists of sappy leaves, e.g. dandelion leaves and petals.

The eggs are greyish brown, oval, and slightly compressed, 5 × 2 mm. They are laid singly in the soil. The total laid varies, but is usually in excess of 50.

The first eggs hatch in May, and nymphs have been observed up to the beginning of August. The nymphs inhabit the same places as the adults. Adults can be found from the beginning of July and on into September. There are probably six instars.

Stridulation. Stridulation in this species consists of a series of powerful, loud chirps, the intervals between which decrease with rising temperature. Each chirp sound like a hoarse, tuneless, abruptly terminated note. Each chirp consists of six syllables (Fig. 17). Stridulation and activity start before noon and continue until late in the afternoon. Stridulation only takes place at temperatures above 23-25°C, which frequently occur at the surface of the ground, where these bush-crickets live (Tetens Nielsen, 1938).

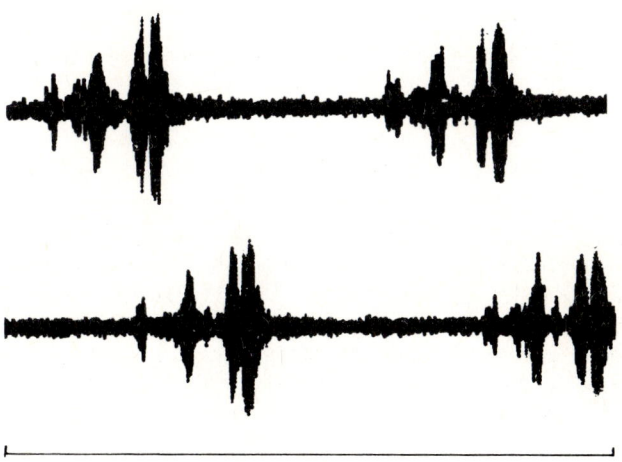

Fig. 17. Oscillogram of the song of *Decticus verrucivorus* (L.). — DK, NEZ: Tisvilde, September 1964. Scale: 0.25 sec.

Genus *Platycleis* Fieber, 1852

Platycleis Fieber, 1852, *in* Kelch: Orthopt. Oberschlesiens: 2.
Type-species: *Locusta grisea* Fabricius, 1781.

Medium-sized bush-crickets. The pronotum is flat anteriorly, the posterior portion having a central keel. No spines on prosternum. In the only species occurring in N. Europe, the wings extend out over the apex of the abdomen. The fore wings, and especially the radial portions, have dark spots bordered by white veins. The movable flaps of the hind tarsi are shorter than the metatarsi. Four spurs on the hind tibiae. Cerci in male conical, straight, with an inward-directed peg on each. Ovipositor slightly to highly upwardly curved.

The genus is divided up into a number of subgenera; the Nordic species belongs to *Platycleis* s. str.

7. *Platycleis (Platycleis) albopunctata* (Goeze, 1758)
Fig. 18.

Tettigonia Albopunctatus Goeze, 1778, Ent. Beytr. 2: 89.
Locusta denticulata Panzer, 1796, Fauna Ins. Germ. 30.
Eng.: Grey Bush-cricket; Dan.: Sandgræshoppen; Fin.: Hietahepokatti; Sw.: Grå vårtbitaren.

Individuals found in Scandinavia are basically brown in colour, interspersed with deeper coloured black and brown patches. The underside of the abdomen is yellowish to greenish in colour. The lateral surface of the hind femora has a darker longitudinal stripe. The fore wings extend beyond the apex of the abdomen but not beyond the end of the hind tibia. The tenth tergum has an arched indentation posteriorly. The subgenital plate in the female is somewhat heart-shaped and frequently slightly grooved. Ovipositor slightly curved. Length: ♂ ♀ 16-23 mm; ovipositor 8-11 mm.

Ander (1949), on the basis of statistical material, demonstrated that there was a series of local forms.

Distribution. In Denmark, this species is common in Tisvilde Hegn and Asserbo Plantage in NEZ. A few have been found on the Refnæs peninsula (NWZ) and on the island of Anholt (EJ). Recorded at various sites on the island of Bornholm. — In Schleswig-Holstein it is common in the area round Möhl, south of Lübeck. — In Sweden, it has been found several times on Kullen in Skåne. It is common on Öland and the Alvar areas of Gotland and, in the rest of Sweden, sporadically throughout the south. — In Norway, recorded in Østfold and Aust-Agder. — In Finland, only on Alandia.

Ander (1949) considers that it immigrated during the Boreal Period, when the climate was drier than today, the original population thereafter being isolated in specially warm and dry places. It is difficult to explain the distribution of this species without accepting it as a relic from the Boreal Period, when Southern Scandinavia was part of the Continent. — The total distribution of the nominal form is sporadic, and extends from S. England and Wales, Spain, Central Europe to Poland and Rumania.

Biology. This is a particularly heat-loving species, remaining close to the ground on low plants (grass, heather, lyme grass etc.) or on cliffs, rocks and bare sand in places exposed to the sun. It is found on beaches, in the vegetation of dunes and moorlands, on Alvar areas, on the fringes of woodlands and in clearings (fire-breaks in pine forests). If disturbed it crawls into the greenery, but it may — to a greater extent than is usual in bush-crickets — elect to fly away. It has been seen to fly for one to two metres (Ander, 1947). It lives off small insects (larvae, aphids etc.) and the soft parts of plants.

The eggs are dark grey, elongated, 4.5 × 1 mm. They may be laid in a variety of places: in crevices in dry stems, treetrunks, etc., or in firm soil protected by a thin layer of moss. The total number of eggs is from 50 to 60.

The first eggs hatch in May (or occasionally at the end of April) and nymphs have been observed up to the end of July. The nymph remains in the same localities as the imago. The imago is found from July to October. Probable number of instars: six.

Stridulation. Stridulation in this species consists of a long series of chirps repeated in a rhythm varying with temperature. Each chirp has a "See-ee-eet" or a "See-ee-ee-eet" sound, with either three or four syllables distinguishable. It stridulates both night and day.

Fig. 18. Oscillogram of the song of *Platycleis albopunctata* (Gz.). — DK, NEZ: Tisvilde, August 1967. One sequence consists of five chirps. Scale: 0.25 sec.

Genus *Metrioptera* Wesmael, 1838

Metrioptera Wesmael, 1838, Bull. Acad. Sci. Bruxelles 5: 592.
  Type-species: *Gryllus brachypterus* Linnaeus, 1761.

A medium-sized cricket. Pronotum flat anteriorly with a central keel posteriorly. Prosternum without spines. In the northern European species, the wings do not extend beyond the apex of the abdomen. Radial portion of fore wings uniform in colour. Movable flaps of hind tarsi shorter than first metatarsus. Four spurs on hind tibiae. Cerci conical in male; straight, and with an inward-facing tooth. Ovipositor variable in length, and more or less sharply upward curving.

The genus is divided into several subgenera.

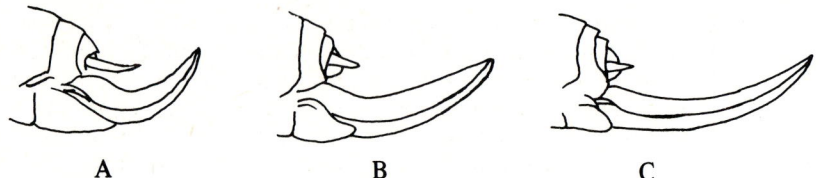

Fig. 19. Side view of the posterior part of the female abdomen of A: *Metrioptera bicolor* (Phil.); B: *M. roeseli* (Hag.); C: *M. brachyptera* (L.).

Key to species of *Metrioptera*
(with indications of subgeneric affinity)

1 Male cerci slender, terminal tooth as long as 20-25% of the entire length (Figs 20A & B). Ovipositor more or less sharply upcurved near the base, at most slightly longer than pronotum (Figs 19A & B). 2

– Male cerci stumpy, terminal tooth at least as long as 33% of the entire length (Fig. 20C). Ovipositor gradually upcurved, nearly twice as long as pronotum (Fig. 19C) . . . . 10. *(Metrioptera s. str.) brachyptera* (Linnaeus)

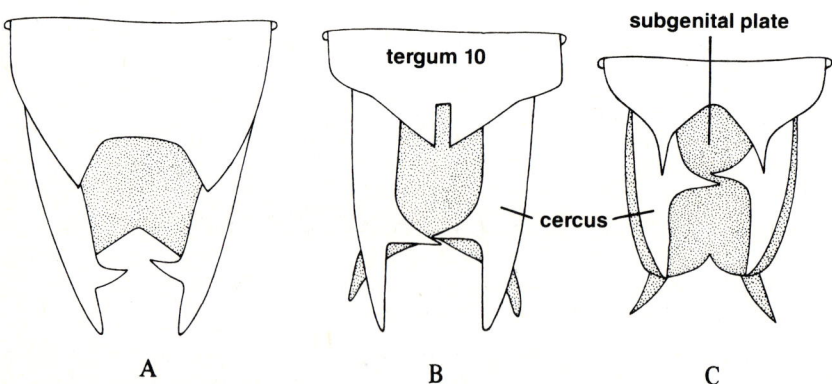

Fig. 20. Dorsal view of the tip of the male abdomen of A: *Metrioptera bicolor* (Phil.); B: *M. roeseli* (Hag.); C: *M. brachyptera* (L.).

2 (1) Paranota with margin more or less pale. Tergum 10 in male
     not cloven (Fig. 20B). Tooth on cercus about a third of the
     way from the apex .... 9. *(Roeseliana) roeseli* (Hagenbach)
–   Paranota all one colour, without a light margin. Tergum 10
     in male deeply cloven (Fig. 20A). Tooth on cercus close to
     apex .... 8. *(Bicolorana) bicolor* (Philippi)

8. *Metrioptera (Bicolorana) bicolor* (Philippi, 1830)
    Figs 19A, 20A, 21.

Locusta bicolor Philippi, 1830, Orth. Berol.: 24.
Sw.: Gröna hedvårtbitaren.

Main coloration: light-green to yellowish. Dorsal surface brownish. Lateral surface of hind femur has a dark, zig-zag, longitudinal stripe. Fore wings in female wide and ovoid, unlike the other species in this genus, whose fore wings are lanceolate. The subgenital plate in the female is uniformly convex, long, and narrow, with only a small incision or notch posteriorly, and with no longitudinal keel. The ovipositor is short, and curves upwards almost at a right angle. The cerci in the male extend a little way beyond the subgenital plate with styli, each cercus having an inward-facing, narrow tooth close to its tip. Tergum 10 in the male deeply cloven posteriorly. Length: ♂ ♀ 12-16 mm; ovipositor 8-10 mm.

Distribution. Not found in Denmark, Finland or Norway. — In Sweden, only reported from localities near the lake Vombsjön, in Skåne. — In Schleswig-Holstein, it has been found at several sites near Mölln, south of Lübeck. — By nature an animal of the steppes, its distribution extends from France, Central Europe, N. Italy and Yugoslavia to Siberia and Mongolia.

Biology. Found in dry, warm places. In Holstein, its habitat comprises places particularly exposed to the sun, where heather, broom or grass grow, or in denser thickets.

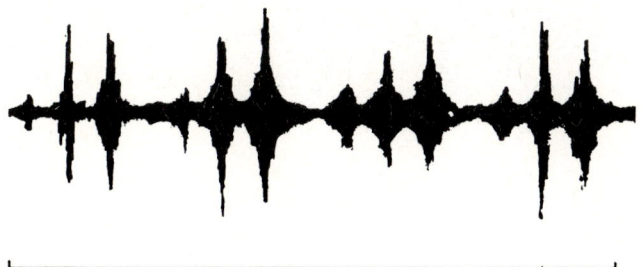

Fig. 21. Oscillogram of the song of *Metrioptera bicolor* (Phil.). — D, Holstein: Göttin, August 1967. Scale: 0.25 sec.

Stridulation. Stridulation consists of a long series of rapid chirps sounding like "deet-deet-deet.." (about fifteen "deets" per second at 20°C), which are just perceptible but difficult to count. The oscillogram (Fig. 21) reveals that each chirp consists of three syllables, which the human ear cannot distinguish apart. Several chirps are produced in sequences of variable duration. At high temperatures, the individual chirps merge.

9. *Metrioptera (Roeseliana) roeseli* (Hagenbach, 1822)
Figs 19B, 20B, 22, 23.

Locusta roeseli Hagenbach, 1822, Symb. Faun. Ins. Helv.: 39.
Eng.: Roesel's Bush-cricket; Fin.: Ruskea töpökatti; Sw.: Ängsvårtbitaren.

Basic colour: yellowish brown or yellowish green, alternating with black and yellow patches. Paranota with sharply delineated light margins, yellow or red. Lateral surface of hind femur bearing a dark, zig-zag, longitudinal stripe. Wings transparent, brown. Subgenital plate in female has a longitudinal keel, and is divided posteriorly for one third of the length of the plate. The ovipositor is only slightly longer than the pronotum and highly, though uniformly, curved upwards. The cerci in the male extend posteriorly a little way beyond the subgenital plate with styli, each cercus having a long, inward-facing tooth one third of the way back from its tip. In addition, tergum 10 in the male is only slightly cloven posteriorly.

Individuals are occasionally found with more highly developed wings than the typi-

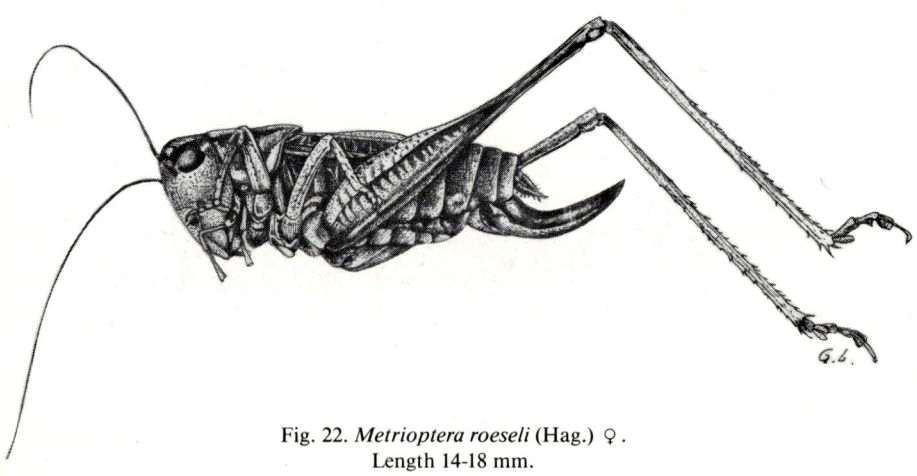

Fig. 22. *Metrioptera roeseli* (Hag.) ♀.
Length 14-18 mm.

cal form. Their fore wings can be longer than the abdomen, or both the fore and hind wings can be much longer than the abdomen. Such specimens have been reported from Falster in Denmark and Borgåstad in Finland. Length: ♂ 12-15 mm; ♀ 14-18 mm; ovipositor 6-8 mm.

Distribution. In the area, only widely distributed and common in the southern provinces of Finland, including Alandia (the Åland Is).). Also round L. Ladoga in Russia. — In the rest of Fennoscandia and Denmark, only recorded in a few localities: near Västerås (Vstm.) in Sweden, and in the western half of Falster and on Lolland (both LFM), and near Herning (WJ) in Denmark. Also several records south of Lübeck in Holstein. — Sporadically in S. and E. England. From France, Belgium, Spain, C. Europe, Yugoslavia and Hungary to Siberia. Introduced to North America (Montreal, Quebec, New York).

Biology. Found in both dry and damp places, in more or less dense thicket. Known from bogs, woodland meadows, "grey" dunes and dry roadsides. It frequently makes for the denser vegetation where the humidity is higher. It is assumed that the long-winged forms are able to fly. They can be very active on hot days.

The eggs are brown and elongated; 4.5 mm in length and 1 mm wide. They are laid within stems. In slim stems the eggs may be laid individually, whilst in thicker ones several are frequently laid at once.

It is possible to observe nymphs from May to July. The imago is seen from July to October. Six instars have been observed.

Stridulation. It is possible to distinguish individual chirps at lower temperatures, the oscillogram (Fig. 23) showing that each one consists of several syllables. They sound like a long series of rapid chirps all tending to merge into one another and impossible to count. At higher temperatures they merge into one long sequence, which sounds like a loud buzzing.

Fig. 23. Oscillogram of the song of *Metrioptera roeseli* (Hag.). — DK, LFM: Marienlyst, August 1967. Scale: 0.25 sec.

10. *Metrioptera (Metrioptera) brachyptera* (Linnaeus, 1761)
Figs 19C, 20C, 24.

Gryllus brachypterus Linnaeus, 1761, Fauna Svecica ed. 2: 237.
Eng.: Bog Bush-cricket; Dan.: Hedegræshoppen; Fin.: Vihreä töpökatti; Sw.: Ljung-vårtbitaren.

The basic colour is dark brown, with green patches, which may be more or less promi-nent or even absent altogether. The vertex, the upper surface of the pronotum and the anterior and posterior margins of the fore wings are usually green. The paranota usually have a light yellow posterior margin, the colour sometimes extending as far forward as the anterior margin. The lateral surface of the hind femur has a dark, fre-quently zig-zag, longitudinal stripe. The subgenital plate in the female is uniformly convex, long, and narrow, with a small incision or notch posteriorly, and no longitudi-nal keel. The ovipositor is almost twice as long as the pronotum, curving slightly up-wards. The cerci in the male are shorter than the subgenital plate, each cercus having an inward-facing tooth in the middle. In addition, tergum 10 in the male is deeply cloven posteriorly.

Forms with well-developed wings arise as a rarity. Length: ♂ ♀ 12-16 mm; oviposi-tor 8-10 mm.

Distribution. In Denmark, it occurs commonly in Jutland, particularly in the West and North, as well as in North Zealand, although it has not been reported from any other of Denmark's islands. — Common in S. Holstein. — In Sweden and Finland it has been recorded in most provinces except the northernmost ones. It has also been found in the contiguous areas of the Soviet Union. — In Norway, it is found in the south, extending no further up the W. coast than to Hordaland. — A palaearctic spe-cies ranging from the British Isles to the Far East of the USSR, south to the Pyrenees and N. Italy, and to Yugoslavia and Rumania.

Biology. This species requires both warmth and humidity, so it prefers localities in which bog vegetation alternates with moorland, allowing it to choose the temperature

Fig. 24. Oscillogram of the song of *Metrioptera brachyptera* (L.). — S, Sm.: Lilhult, September 1979. Scale: 0.25 sec.

and humidity conditions best suited to its requirements. Common in woodland meadows. Can be very active on hot days. Feeds on small insects and soft parts of plants.

The eggs are dark brown and slightly oval in shape, 4×1 mm. They are laid in the soil and on plants.

The first eggs hatch in May, nymphs being observed until the beginning of August. The nymph inhabits the same sites as the imago. Imago found from July to the end of August, and rarely right on into October. Six instars have been observed.

Stridulation. Stridulation in this species consists of a long series of chirps which may be repeated at a variable rhythm depending on temperature. Each chirp sounds like "Sseer", with three distinct syllables. Stridulation takes place both by day and by night.

Genus *Pholidoptera* Wesmael, 1838

Pholidoptera Wesmael, 1838, Bull. Acad. Bruxelles 5: 592.
Type-species: see Harz, 1969: 325.

Medium-sized bush-crickets with short scale-like fore wings. Pronotum smooth, with rounded sides (central keel found in rare cases). The scale-like fore wings are partially concealed by the rear margin of the pronotum. The fore wings in the female can just be made out behind the rear margin of the pronotum, whilst those in the male extend further out. No spines on the prosternum. The flaps on the hind tarsi are only slightly shorter than the first segment of the tarsus. Four spurs on the hind tibiae. The cerci in the male are conical, straight, and each equipped with a small, inward-facing spine. The ovipositor is non-denticulated and from slightly to highly curved.

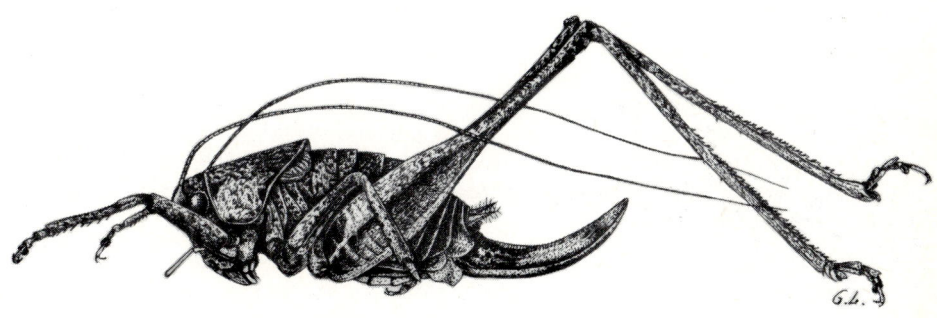

Fig. 25. *Pholidoptera griseoaptera* (De Geer) ♀.
Length 13-20 mm.

11. *Pholidoptera griseoaptera* (De Geer, 1773)
Figs 25,26.

Locusta griseoaptera De Geer, 1773, Mém. Ins. 3: 436.
Eng.: Dark Bush-cricket; Dan.: Buskgræshoppen; Fin.: Pensashepokatti; Norw.: Buskhopper; Sw.: Buskvårtbitaren.

The basic coloration is brownish, though the hues may vary somewhat on the border-lines to reddish and black. The head is marbled. The vertex and the upper surface of the pronotum are a lighter shade of brown. Upper surface of pronotum flat or slightly convex. The paranota frequently have a darker patch towards the rear, with a fine yellow line on the margin. The legs are marbled. The lateral surface of the hind femur has a black longitudinal band. Lower surface of abdomen yellowish. Ovipositor sharply curved upward. Subgenital plate in male folded along the midline, thus coming to comprise two rounded lobes. The cerci in the male each have an inward-facing spine near the base. Length: ♂ ♀ 13-20 mm; ovipositor 9-11 mm.

Distribution. In Denmark, this insect occurs commonly in the E. of Jutland and on the Islands, but it does not seem to occur in WJ or NWJ. In NEJ, only at Blokhus. — In Schleswig-Holstein it has been observed in the eastern and southern regions. — Sweden: extends as far north as Värmland and Uppland. — In Norway, it has been recorded round Oslo Fjord and in isolated sites near the coast towards Kristiansand. — In Finland: on Alandia and in the southernmost provinces, and on into the Soviet Union. — Recorded from England and Wales, France, C. Europe, N. Spain, N. Italy, Yugoslavia; in the USSR to the Urals.

Biology. Remains close to the ground, and can be found in gardens, parks, roadside verges, hedges, the fringes of and clearings in woods, and in general thicket scrub. It inhabits places where the light is not too bright, remaining in the low vegetation in particular, rarely climbing more than two metres up. The diet consists of small insects and the soft parts of plants.

The eggs are light brown in colour, and faintly oval in shape, 3-4.5 mm. They are laid in cracks in bark and on rotten wood late in the summer and on into October. The eggs evolve slightly the first year, continuing their development the following summer and, when the embryo is fully developed about October, it enters a period of diapause. The eggs do not hatch until the following (third) year, in May (Hartley & Warne, 1973). The nymph remains more at the surface of the soil than the imago. The first adults are found in July, and they can commonly be heard stridulating until late on in October. It is probable that there are six instars.

Stridulation. Stridulation consists of a series of chirps with a "sseet" sound, repeated at a variable rhythm depending on temperature. At low temperatures each chirp becomes a slightly longer "sseeet". The chirp can be heard several metres off. Stridulation — and thereby also activity — commences on into the afternoon and may be kept up all night if the temperature is high enough. If a male is brought into a dark room in the middle of the day, it will immediately begin to stridulate. Stridulation is set off by fading daylight and ceases when the temperature gets down to 5-7°C. Each chirp consists of four syllables (Fig. 26).

Ph. griseoaptera lives close to the surface of the soil, unlike *T. viridissima,* which lives higher up and where lighting and temperature are also of significance in stridulation. If two males happen to be sited close to each other they commence "rivalry singing", presumably to demarcate their territories. Each chirp normally lasts for about a tenth of a second at 14°C (Fig. 26A), but can last up to a fifth of a second at 8°C (Fig. 26B). It can last up to two fifths of a second in "rivalry singing" (Jones, 1966).

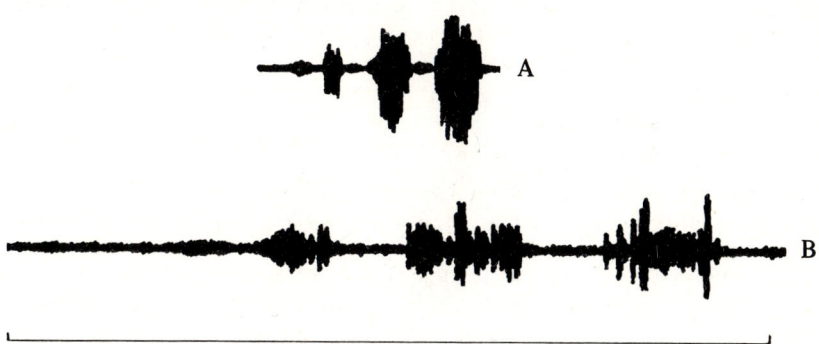

Fig. 26. Oscillogram of the song of *Pholidoptera griseoaptera* (De Geer). — DK, NEZ: Ryget Skov, September 1980. The same specimen is recorded twice, above at 14°C, below at 8°C; each row shows one chirp. Scale 0.25 sec.

Superfamily Gryllacridoidea

Antennae long and thin. The tarsus consists of four segments. No tympanal organs. No stridulatory organs, even in tropical species with highly developed wings. Cerci long and flexible in both sexes. Styli generally present in males. Ovipositor long, usually comprising three pairs of valves.

Family Rhaphidophoridae

Always wingless. Legs, antennae and palps very long. Tarsus elongated and compressed. Two small triangular tubercles on the vertex between the eyes.

Brownish or yellowish species, found particularly in caves, shady forest, under leaves and stones, and in greenhouses. Several free-living species in S. Europe.

Genus *Tachycines* Adelung, 1902

Tachycines Adelung, 1902, Ann. Mus. Zool. Ac. Sc. St. Pétersbg. 7: 56.
 Type-species: *Tachycines asynamorus* Adelung, 1902.

Fastigium grooved dorsally, apex projecting as two horns. Antenna up to 4 times as long as body. Pronotum convex dorsally. Fore femora with a long movable spine on outer side. Metatarsus of hind leg as long as other segments together.
 One cosmopolitan species.

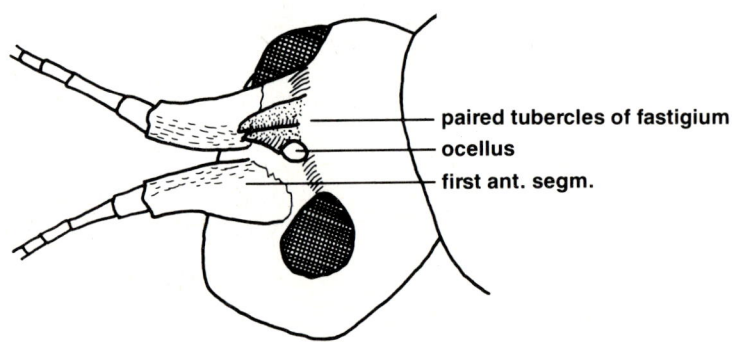

paired tubercles of fastigium
ocellus
first ant. segm.

Fig. 27. Head from above of *Tachycines asynamorus* Ad.

12. *Tachycines asynamorus* Adelung, 1902
Figs 27, 28.

Tachycines asynamorus Adelung, 1902, Ann. Mus. Zool. Ac. Sc. St. Petersbg. 7: 59.
Eng.: Greenhouse Camel-cricket; Dan.: Væksthusgræshoppen; Fin.: Ansarihepokatti; Norw.: Veksthusgrashoppe; Sw.: Växthusgräshoppan.

Basic colour yellowish brown, interspersed with dark and light patches. Cerci nearly 10 mm long in both sexes, rather upwardly curved toward the tip. No styli in the male. Ovipositor practically straight; only very slightly upwardly curved.
 Length: ♂ ♀ 12-17 mm; ovipositor 11-12 mm.

Distribution. This species is considered to be a native of China, from whence it was introduced into Europe at the end of the Nineteenth Century (Prague 1891, Hamburg 1892, St. Petersburg 1902, at which place it was first described). It has now spread practically throughout Europe, where it can be quite common in greenhouses. Cosmopolitan.

Biology. As the name indicates, the Greenhouse Camel-cricket lives in greenhouses, conservatories and botanical gardens. It may be met with out of doors in the spring and summer, but it cannot survive the winter in the open. It is definitely an insect of the dusk and the dark, going into hiding by day. It is occasionally mistaken for a spider, because of its long legs. It can jump up to 1½ metres. It lives largely on animal food, e.g. aphids. It also consumes carrion and members of its own species. It does, however, also eat plant material, in the event of their being a lack of animal food. In order to develop properly, it requires both animal and plant food in its diet, as well as water. There can be no doubt that it is of some benefit in greenhouses, devouring parasites like aphids, but it may on the other hand occur in such large numbers that it damages the plants such as seedlings and Chrysanthemum flowers.

No form of sound production known.

Eggs white and rather oval, 2-2.5 × 1.2 mm. The eggs are laid in the soil individually or in batches of small numbers. They take from two to four months to hatch, depending on temperature. Laying begins shortly after the final moult. A female has been observed to lay up to 50 eggs in one night. One single female may lay up to a thousand eggs or more in the course of her life, depending on temperature.

Nymphal development takes from four to seven months, also depending on temperature. Since no wings develop, it can be a little difficult to tell the various instars apart, but an indicator in this direction in the female is the length of the ovipositor. Ten instars have been observed.

The Greenhouse Camel-cricket is easy to keep in captivity and is very suitable for laboratory experiments.

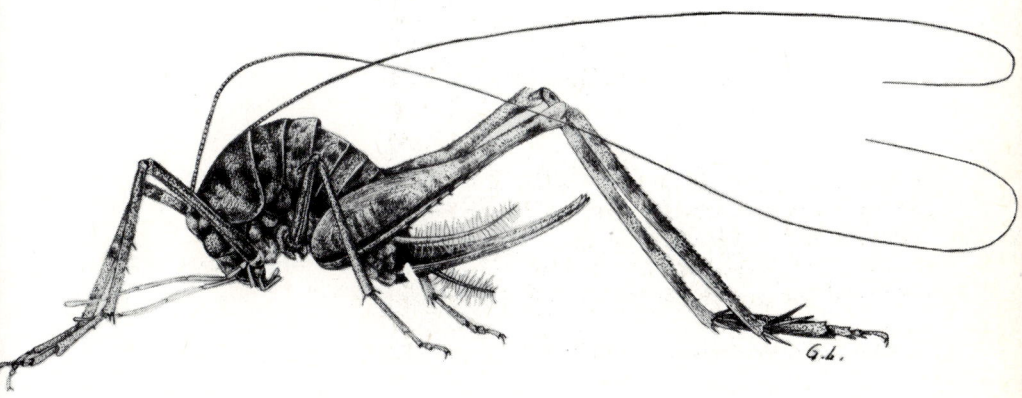

Fig. 28. *Tachycines asynamorus* Ad. ♀.
Length: 12-17 mm.

Superfamily Grylloidea

The body is more or less cylindrical (though also oval in southern species). The antennae are long and filiform, except in the highly specialised Gryllotalpidae, where they are brush-shaped. Pronotum with neither central nor lateral keel. Sternal plates flat, with no spines or flaps. Tarsus consists of three segments. Fore tibia usually bears the tympanal organ, where the tympana are visible. In the two species of cricket found in N. Europe, the fore tibia has two tympana: a small one medially and a large one laterally, whilst the mole cricket has only one, covered tympanal opening on the medial tibial surface. The postcubital section of the fore wings (adjacent to the back) is large and occupied by the stridulatory organ (Fig. 5B). There are both plectrum and pars stridens on the right and left fore wings. The postcubital section of the right fore wing overlaps that of the left. The cerci are long and flexible. No styli on the subgenital plate. The ovipositor is well developed in Gryllotalpidae. It comprises two pairs of valves.

Key to families of Grylloidea

1　Antennae short. Fore legs highly specialised, modified for
　　digging. Pronotum highly enlarged .... Gryllotalpidae (p. 56)
－　Antennae long. Fore legs and pronotum normal .... Gryllidae (p. 50)

Family Gryllidae

(Crickets)

Medium-sized to small insects. Head spherical. Antennae long and filiform. Ovipositor usually long, narrow, cylindrical, and smooth and shiny. Hind legs usually with enlarged femora for jumping.

This family is subdivided into a number of subfamilies which will, however, not be discussed here. The two genera occurring in Scandinavia belong to the subfamily Gryllinae.

About 70 species are known in Europe.

Key to genera of Gryllidae

1　Body black and shiny, practically hairless .... *Gryllus* Linnaeus (p. 52)
－　Body yellowish to brownish, finely and densely hairy .. *Acheta* Fabricius (p. 53)

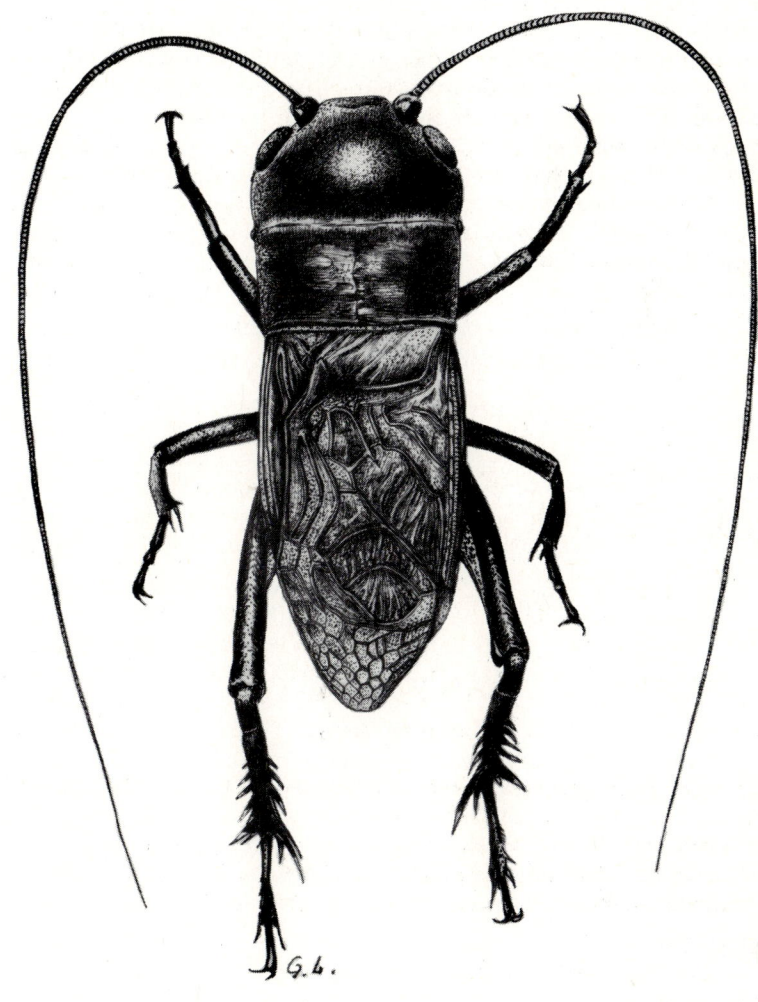

Fig. 29. *Gryllus campestris* L. ♂.
Length 17-26 mm.

Genus *Gryllus* Linnaeus, 1758

Gryllus Linnaeus, 1758; Syst. Nat. ed. 10: 425.
  Type-species: *Gryllus campestris* Linnaeus, 1758.

A strongly built and compact species. The three ocelli arranged nearly in a straight line. Pronotum wider before than behind. Fore tibia with two tympana. Hind femora powerful; as long as or a little shorter than hind tibia and tarsus together. Metatarsus highly spiny.
  Two species in Europe; only one in N.Europe.

13. *Gryllus campestris* Linnaeus, 1758
  Fig. 29.

Gryllus campestris Linnaeus, 1758, Syst. Nat. ed. 10: 428.
Eng.: Field-cricket; Dan.: Markfårekyllingen; Fin.: Kenttäsirkka; Sw.: Fältsyrsan.

The basic colour is black and — as a result of the comparative lack of hairs — shiny as well. The fore wings are brownish, though yellowish towards the base. Hind femora red underneath and on the medial surface. Head large, and wider than the pronotum — particularly in the male. Pronotum slightly wider anteriorly than posteriorly. Fore wings extend to the apex of the abdomen, the hind wings being shorter than the fore wings. Length: ♂ ♀ 17-26 mm; ovipositor 12-18 mm.

Distribution. In Denmark, the field cricket has only been found on Bornholm. The oldest find was recorded in 1897, at Galløkken, near Rønne. In 1947, this cricket was reported at several sites along the south coast, from a little to the west of Rønne airport and to Dueodde. It is presumed that the insect inhabited the island from 1943 and up to about 1957 (Larsen, 1944). There is a considerable amount of disagreement about whether the field cricket has been introduced to the island from time to time from the east — the most recent occasion being by Russian troops in 1945 (after the War) — or whether there is a small permanent population which increased widely between 1943 and 1957.
  Not recorded in Sweden, Norway or Finland. — In Holstein it has been found near Lübeck and in several localities S. of there, near Ratzeburg. — The total distribution ranges from S.England to West Asia and North Africa.

Biology. The field cricket is met with especially in hot, dry areas on moorlands, in dunes, in clearings in woods, on sandy meadows and on south-facing slopes. This species lives in holes which it digs itself, the male frequently sitting in the entrance, stridulating. If alarmed, it rapidly retreats into its hole. It is frequently possible to catch one by poking a stem of grass down the hole and quickly pulling it out again. The cricket will often come up with the grass, having bitten this disturber of its peace. The field cricket is omnivorous.

Eggs are laid from May to July, a single female being capable of laying several hundred eggs. They hatch in from two to four weeks, the nymphs increasing in size during the summer. The adults die during the course of the summer. It seems to be the tenth instar which passes the winter; prior to hibernation the well developed nymph digs a hole underground. To do this, it uses its fore limbs and its very large mandibles as "tools", the loose earth being scraped out backwards between the legs. The tunnel is commonly 30 to 40 cm long, and can go down to a depth of 30 cm. The final moult takes place in the spring. Whilst the adult male frequently remains in the hole, the female sets off on a search for a "singing" male. The female frequently remains in the male's hole, even after the first copulation. The female often lays her eggs in the soil immediately outside the hole. Eleven instars have been observed.

Stridulation. Stridulation is very loud, usually consisting of three to four chirps per second. Each chirp comprises three or four syllables, which cannot be distinguished apart by the human ear. In hot weather the field cricket will sing (stridulate) practically round the clock, but for a short pause about sunrise. In cold weather, stridulation ceases. The male commences its stridulatory call in early May, and it may last until the end of June. Some may, indeed, continue on into July.

Genus *Acheta* Fabricius, 1775

Acheta Fabricius, 1775, Syst. Ent.: 279.
Type-species: *Gryllus domesticus* Linnaeus, 1758.

Similar to *Gryllus,* but usually paler in colour, not glossy, and densely covered with hairs. Pronotum same width throughout.
The genus is cosmopolitan; only three species are known in Europe.

14. ***Acheta domestica*** (Linnaeus, 1758)
Figs 5B, 30, 31.

Gryllus domesticus Linnaeus, 1758, Syst. Nat. ed. 10: 428.
Eng.: House-cricket; Dan.: Husfårekyllingen; Fin.: Kotisirkka; Norw.: Hus-siriss; Sw.: Hussyrsan.

Basically yellowish in colour, with several brown patches, especially on the head and pronotum. Two brownish coloured transverse grooves across the vertex: one between the eyes and another further back, near the pronotum. The frons is largely brownish, with patches of lighter colour scattered about; there are brown markings superiorly on the pronotum. The fore wings extend nearly to the apex of the abdomen; rarely beyond it. The hind wings extend far beyond the apex of the abdomen, where they are folded together in such a way as to resemble two cords. The ovipositor is long. Length: ♂ ♀ 16-20 mm; ovipositor 11-15 mm.

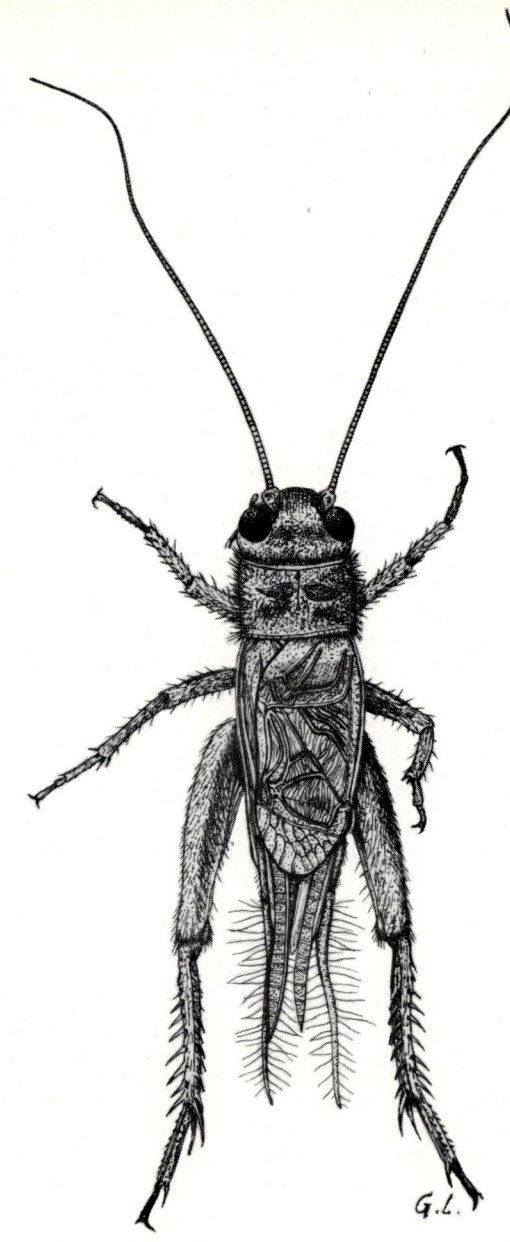

Fig. 30. *Acheta domestica* (L.) ♂.
Length 16-20 mm.

Distribution. Cosmopolitan. It is presumed to be a native of North Africa and the Near East, where it lives quite naturally and from which it has been introduced to Europe, including Denmark and Fennoscandia.

Biology. The house cricket is found indoors in houses, bakeries, restaurants, greenhouses etc. and, in the outdoors, on rubbish dumps in the southern parts of North Europe, where it may be very common. Its movements are very quick — quicker than those of the field cricket — and it can be difficult to catch. It prefers to jump to safety rather than to fly. It can occasionally be seen out of doors in summer, close to human habitation — especially in hot summers. It eats practically anything (flour, bread and other kitchen waste; cloth, paper, leather, dead animals etc.). It is quite capable of digging a hole in the ground, just like the field cricket, but indoors it usually confines itself to cracks and crevices. Its habits are decidedly nocturnal.

Whilst the house crickets which live inside are independent of the seasons for their development, the ones living outside will to a certain extent be dependent on ambient temperatures. On rubbish dumps, they appear to be capable of hibernating as eggs, and even nymphs and adults have been observed in the winter months when temperatures were below freezing point, since they live deep in the rubbish at levels where the processes of fermentation keep the temperatures up.

A female will commonly lay some two to three hundred eggs. They may be laid in the soil or in cracks between floorboards, in crevices, etc. Development and life-span are highly temperature-dependent. The number of instars may vary: 7 to 13 have been confirmed.

Stridulation. Stridulation in the house cricket is not quite as loud as in the field cricket, consisting of from one to several chirps per second, depending on temperature. Each chirp consists of three or four syllables, though the human ear cannot distinguish them apart. As a rule, they stridulate in the evening and at night, but they have also been heard "singing" by day on rubbish dumps in very hot weather.

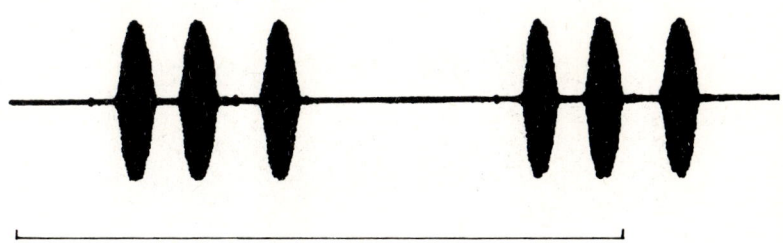

Fig. 31. Oscillogram of the song of *Acheta domestica* (L.). — DK, NEZ: Ledøje, June 1968. Scale: 0.5 sec.

Family Gryllotalpidae

(Mole crickets)

Generally very large insects. The head is virtually spherical, with prognathous mouthparts. Eyes present. The antennae are short but multi-segmented, as in all Ensifera. The pronotum is highly expanded, and the paranota completely enclose the prothorax. In this way, the pronotum wraps round the prothorax like a cylinder except for the inferior aspect, where the powerful, closely placed digging limbs originate. The fore legs are adapted for digging, assisted by the powerful musculature of the prothorax. The hind legs are not modified for jumping but for running. Fore tibia with one tympanum on inside. Hind wings generally longer than fore wings. No ovipositor.

As can be deduced from all the above characteristics, this group is above all adapted to a subterranean, excavatory existence. This adaptation is, however, not so exclusive as to prevent the mole cricket moving about above ground, both on the surface and in the air.

One genus (four species) known in Europe.

Genus *Gryllotalpa* Latreille, 1802

Gryllotalpa Latreille, 1802, Hist.Nat.Crust.Ins. 3: 275.
  Type-species: *Gryllus gryllotalpa* Linnaeus, 1758.

With the characters of the family.

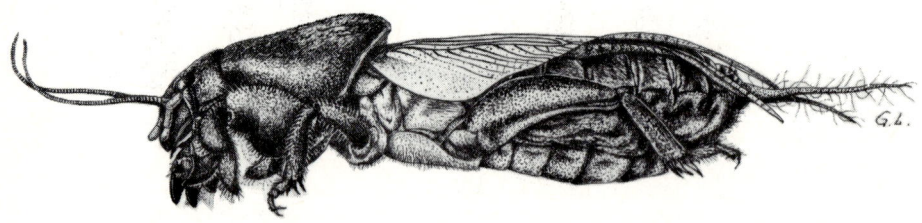

Fig. 32. *Gryllotalpa gryllotalpa* (L.). Length 35-40 mm.

15. *Gryllotalpa gryllotalpa* (Linnaeus, 1758)
Fig. 32.

Gryllus gryllotalpa Linnaeus, 1758, Syst.Nat.ed.10: 428.
Eng.: Mole-cricket; Dan.: Jordkrebs; Fin.: Maamyyräsirkka; Norw.: Jordsiriss; Sw.: Mullvadssyrsan.

Brownish in colour, the body being covered in a dense felted mass of hairs. Fore wings short. Hind wings extend a short way beyond the apex of the abdomen and are folded beyond the fore wings, resembling two cords.

The female not having an ovipositor makes it difficult to distinguish between the sexes. One of the ways of telling the male from the female is that it has a well developed stridulatory organ.

Length: ♂ ♀ 35-50 mm.

Distribution. Denmark: only found in the east of the country, and in recent years found particularly commonly on Zealand and Møn. — Observed at several places in Schleswig-Holstein, particularly in the south. — Found several times in S. Sweden: Sk. to Öl. and Vg. — Not in Norway or Finland. — The total distribution ranges from the British Isles to W.Asia and N.Africa.

Biology. The sites primarily favoured by the mole cricket are damp places in meadows and near streams, lakes and bogs. It may also be found in heavily watered gardens and parks, in hotbeds, frames and greenhouses. It has also been found in roadside ditches and in drainage ditches in fields. It usually keeps to the criss-cross tunnels it makes underground, their depth probably depending on the humidity of the soil. It is also capable of movement above ground, both running and flying: it can in fact run quite fast. Its flight is noisy and not a little clumsy, and usually quite near the ground.

Copulation usually takes place in N. Europe from May to July, and possibly into August. The eggs are laid the same year and evolve, with no diapause, in the course of a few weeks, after which they hatch. It seems that the nymphs take two years to develop up here in the North, though the process only takes a year further South in Europe.

The eggs are about 2 mm in length, and are reddish-yellow in colour. They are laid in a little "pod" whose depth underground depends on the humidity. The eggs are laid on the floor of the chamber over a space of a few weeks, up to a total of some one to three hundred.

The nymphs will also occasionally appear above ground level.

Stridulation. Stridulation is very loud, sounding like a permanent sort of buzzing. During stridulation, the male places himself at the mouth of the burrow with his hind quarters on display. Stridulation is especially heard in the late afternoon and evening.

SUBORDER CAELIFERA

Head

The head (Fig. 34) is very broad and large, containing the powerful musculature of the mandibles. In principle, the mouth parts are constructed as in the Ensifera. In the Caelifera, however, they are adapted for the mastication of plant material and not — as in the Ensifera — for grasping and cutting. The fastigia frontis and verticis are usually fused into a wide, protuberant keel. The sides of the fastigium verticis commonly bear a pair of foveolae, which may vary in form from triangular or quadrangular via pentagonal to oval.

The compound eyes are quite large and irregularly oval in shape. The ocelli are three in number, the lateral pair being situated close to the medial border of the eyes.

A feature characteristic of the Caelifera is their short antennae, consisting of only a few segments ('short-horns'). The antennae are usually filiform, but may be other shapes. The Northern European grasshoppers include species with clublike antennae, e.g. *Myrmeleotettix*.

Thorax, legs and wings

The pronotum is constructed in approximately the same way as described under the Ensifera. A median keel is frequently evolved, and it may be more or less prominent, rising to a sharply dome-shaped ridge in some groups. The median keel may be entire or subdivided by up to three transverse sulci, the posterior one being the most prominent. This divides the pronotum into two nearly equal sized portions: a prozona anteriorly and a metazona posteriorly. At the transition from the upper portion of the pronotum and the paranota there is a side keel on each side, which may be straight, or bent at an angle, as shown in figure 34. The pronotum in the Caelifera may adopt some

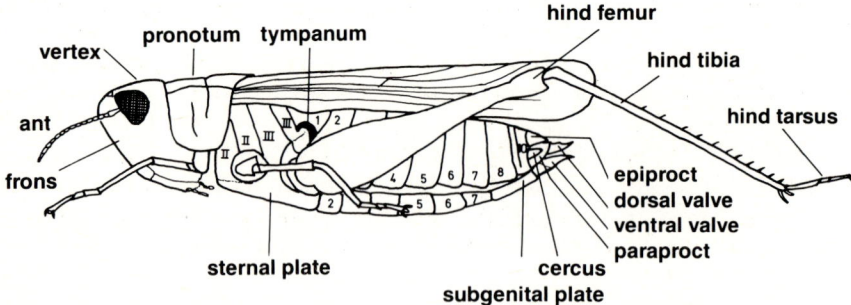

Fig. 33. External morphology of a female grasshopper. — II = mesothorax; III = metathorax; 1-8 = abdominal segments.

quite bizarre shapes, as in e.g. the small *Tetrix* species (ground-hoppers), where its rear margin extends a long way out over the abdomen.

The width of the sterna is considerable, the coxae being widely separated. Together with the prosternum, the mesosternum and metasternum form a heavily sclerotised sternal plate divided by sutures into a row of more or less distinct plates. As described under the Ensifera, the mesosternum and metasternum in the Caelifera also carry a pair of platelike structures. In this suborder, however, they are sunk into the sclerotised plate as lobes.

By and large the legs are constructed as described under the Ensifera.

The wings are generally well developed, the members of the Caelifera in general being better fliers than those of the Ensifera. The ability to fly varies widely in this group, from the locusts, which are long-distance fliers able to remain in the air for hours at a time, to the small grasshoppers of Scandinavia which, even in favourable circumstances on a hot day, never fly far. A few examples of reduced wings are seen amongst Northern European species. As can be seen from the venation, the part primarily reduced is the distal area of the fore wing. A reduction and simplification of the wing venation also takes place at the same time.

The fore wings (tegmina) are also generally long, narrow and membranous, with only a small precostal area. The venation of the fore wings (Fig. 38) is of great significance in identification.

At rest, the hind wings (alea) are kept folded up under the fore wings. They are thin and transparent, with the same venation as the fore wings, plus a total of five extra veins.

Abdomen

The abdomen is practically always large and broadly attached to the thorax. It consists of ten terga, the tenth carrying a small pair of conical, unsegmented cerci in Scandinavian species. The male has nine sterna, the ninth being the subgenital plate, which is

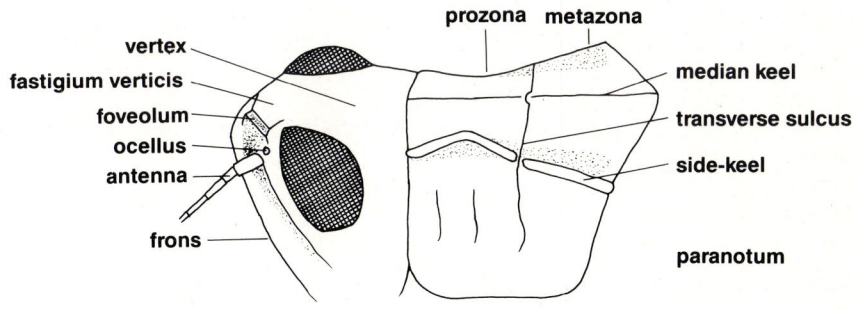

Fig. 34. Head and pronotum of a grasshopper.

powerfully developed and curved upwards. The female has only eight sterna, the eighth forming the subgenital plate, which is much longer than it is wide. There are no styli. The first segment of the abdomen is particularly interesting, because a tympanal (auditory) organ is found on each side, outwardly visible as a more or less restricted or unrestricted tympanum (Fig. 35).

The anus is placed dorsally on the tenth segment, surrounded by three valves: — an upper epiproct and two lower paraprocts. The upper valve is attached as an extension of the terminal dorsal plate, situated like a flap or lobe between the two short cerci (when viewed from above). The two lower valves are beneath this and to the side, each somewhat covered by a cercus.

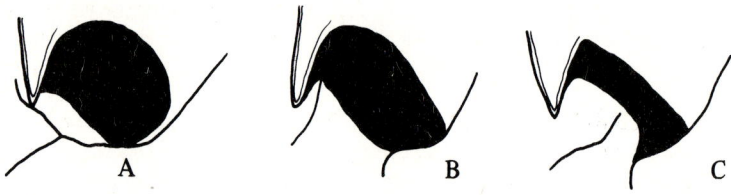

Fig. 35. Left tympanum of grasshoppers. — A: *Mecostethus grossus* (L.); B: *Chorthippus apricarius* (L.); C: *Chorthippus brunneus* (Thnbg.).

The male copulatory organ consists of both a penis and a series of membranous and sclerotised parts (the phallus complex). The penis itself consists of a number of thick sclerotised formations, and is highly curved. It is sited behind the subgenital plate, under a thin pallium found dorsally at the very rear of the plate (Fig. 36). The penis can be seen by pulling the subgenital plate and pallium back. The posterior portion of the penis is supported by a series of more or less sclerotised parts. The entire phallus complex, comprising the penis plus the sclerotised parts, can easily be removed with a fine needle. Superiorly dorsally is the epiphallus, a heavily sclerotised structure which may be plate-like or bridge-like, and which is of systematic significance.

The ovipositor originates on abdominal segments 8 and 9 and consists of three types of valves: — an upper, a lower and an inner pair. This latter pair is rudimentary and is found at the base, between the dorsal and the ventral valves, being invisible from the exterior.

Stridulatory organs

The stridulatory organ in Caelifera, if found, has two components: — a scraper, comprising a prominent, hardened ridge; and a file, consisting of a series of »pegs«. Stridulation is brought about by rubbing the scraper over the file, producing vibrations. The siting of the scraper and file may vary. In the Northern European grasshoppers they are associated with the inner edges of the hind femora and the upper portion of

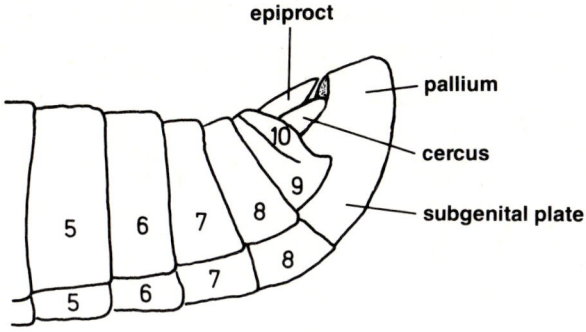

Fig. 36. Side view of the posterior part of the abdomen of a male grasshopper.

the fore wings (tegmina). Stridulation is effected by moving the hind legs up and down, rubbing the hind femora against the fore wings at rest. The latter in consequence vibrate. In the family Acrididae, two groups can be distinguished according to the position of the scraper and file.

In the Gomphocerinae, the radius of the fore wings acts as the scraper whilst the file is on the inner edge of the hind femur (Fig. 37). The fore wings are rooflike, the radius forming the ridge of the roof. The ratio of the length of the file to that of the hind femur may vary. In some genera it amounts to two thirds of the overall length of the hind femur and in others only one third. The number of pegs in the file and the distance between them also varies from species to species. These characteristic features are frequently used to confirm an identification, especially in the closely related species of the genus *Chorthippus,* where the file often has to be studied very closely. This is best done at X 50 magnification. Both males and females are able to stridulate, though the sound is somewhat weaker in the female. The number of pegs in the two sexes is approximately the same within any one species, although the pegs are very much smaller in the female than in the male. Distances between individual pegs vary along the length of the file, their density being greatest in the basal portion (b). Geographical variations in the structure and number of pegs occur within certain species. The numbers of pegs quoted for the various species refer to Nordic individuals, provided the material was extensive enough.

The Locustinae have a stridulatory organ which can in many ways be described as the opposite of that found in the Gomphocerinae, the scraper here being a sharp edge on the inner edge of the hind femur and the file being found on the vena intercalata in the medial area (Fig. 38), which is not normally present in Gomphocerinae. The file comprises a series of uniform "pegs". This stridulatory organ is no absolutely characteristic identificatory feature of the Locustinae, some species having no stridulatory organ and certain genera having totally different stridulatory mechanisms.

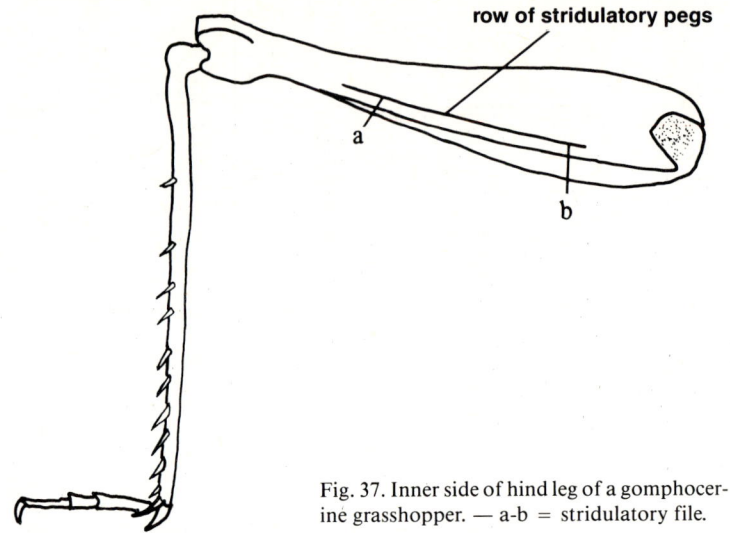

row of stridulatory pegs

a

b

Fig. 37. Inner side of hind leg of a gomphocerine grasshopper. — a-b = stridulatory file.

Development

The vermiform nymph resembles the corresponding stage in the Ensifera. It is surrounded by a thin, transparent membrane, which ruptures when it reaches the surface
of the soil. The vermiform nymphs perform worm-like movements in order to emerge.
It has been agreed to call the first instar larva the phase that emerges on rupture of the
nymphal membrane.

The number of instars may vary, but seems to be constant for any individual species.
In the commonly occurring genera *Chorthippus* and *Omocestus* the lowest number of
instars is four. A key to the British species can be found in Richards & Waloff (1954).
Higher numbers of instars occur in other species. The number of instars varies between the two sexes in *Tetrix,* the male having five and the female six. A key to the British species has been published by Farrow (1964).

The times given for hatching and the emergence of imagines in the autumn are based
on Scandinavian individuals.

The eggs are cylindrical to oval and are laid in ranks in a pod, which can be buried
in the soil or deposited between plants. The female uses the valves of the ovipositor to
dig with, pushing them down into the soil folded together, after which she opens them,
displacing the soil. The abdomen becomes gradually deeper buried, at the same time
being longer and longer. In Gomphocerine species each pod contains 6 to 15 eggs.
Each female may construct several pods. During egg-laying, a frothy substance is
secreted by certain accessory glands in the abdomen and this sets round the eggs. The

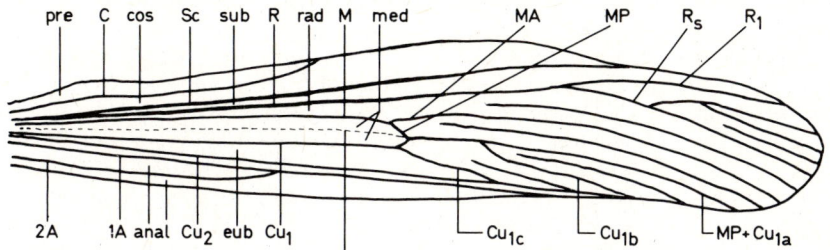

Fig. 38. Left fore wing of a male grasshopper. — C = costa; Sc = subcosta; R = radius; R_1 = first radial vein; R_s = radial sector; M = media; MA = anterior media; MP = posterior media; Cu_1 = first cubital vein; Cu_2 = second cubital vein; 1A = first anal vein; 2A = second anal vein. Also indicated are the principal areas of the wing: pre = precostal area; cos = costal area; sub = subcostal area; rad = radial area; med = medial area; an = anal area. The broken line indicate position of Vena intercalata as found in the Locustinae.

appearance of the pod varies from species to species. The eggs may be laid at an angle to or parallel with the long axis of the pod. The outer wall may consist of hardened froth alone; froth plus fragments of plant material, or froth plus soil particles. There may or may not be a lid at the top of the pod or capsule. The internal structure may be a soft or frothy secretion which may in some cases form lamellae round the eggs. These features may be used in identification, in addition to the structure of the egg itself (Richards & Waloff, 1954). In *Tetrix* the eggs are laid in groups held together by a secretion, but there is no actual capsule or pod.

Grasshoppers pass the winter in the egg, which hatch in May. The last imagines are observed on into October, or more rarely November. Ground-hoppers pass the winter as larvae.

Habitat preferences

All species belonging to the Caelifera are found close to the surface of the soil on low vegetation or on the bare ground, and only in places reached by the sun. They are frequently found in large numbers, also in species occurring rarely but which are common in a restricted area. There is frequently a clear correlation between the composition of the grasshopper population and the composition of the local flora. The individual species' requirements for warmth and humidity are also of significance to their local distribution. On sloping hillsides it is usual to find *Chorthippus parallelus* at the bottom, possibly together with *Omocestus viridulus,* commonly found in damp places. *Ch. brunneus* occurs higher up the slope, with *Ch. biguttulus* and/or the rarer *Ch. mollis* — both species demanding the lowest humidities — at the top. The zones occupied by the various species may move up and down the slope, all depending on whether it is a dry summer or a wet one, since grasshoppers migrate over short distances.

The following species may be found by lakes and streams and in bogs and meadows:- *Tetrix subulata, Mecostethus grossus, Omocestus viridulus, Chorthippus parallelus* and *Chrysochraon dispar.* At the edges of woods and in clearings:- *Tetrix undulata, Omocestus viridulus, O. ventralis* and *Chorthippus brunneus.* Beside ditches and in other permanent stands of grass:- *O. viridulus, Ch. brunneus, Ch. biguttulus, Ch. apricarius* and *Ch. parallelus.* On heaths and moorlands, places with light soil, and on "alvar" areas:- *Bryodema tuberculata* (Jutland, Öland), *Myrmeleotettix maculatus, O. viridulus, O. haemorrhoidalis* (the Mols peninsula (EJ) and Öland), *Gomphocerus rufus, Ch. brunneus, Psophus stridulus* and *Sphingonotus coerulans.* In the "grey" and "white" dunes we may find e.g.:- *Ch. brunneus, Ch. vagans* (at The Skaw (NEJ)), *Ch. albomarginatus* and *Myrmeleotettix maculatus.* In the mountains we find *Melanoplus frigidus* and, just below the tree-line, *Podisma pedestris. Locusta migratoria* has occasionally migrated up from the south-east, but this has been a very rare occurrence in recent years.

Diet

The diet consists of plant material. An investigation carried out in some of the commoner species of the subfamily Gomphocerinae has shown that they prefer grass. The blades are attacked from the side, with rapid mastication. *Tetrix* frequently eats algae, lichen and moss.

Colours

The colours may in some cases vary widely (see under the individual species). The pattern of distribution of colour appears in some cases to be hereditary, since green and brown individuals collected in the same locality and kept in captivity retain their colour for a couple of generations at least, regardless of the colour of their surroundings. The colour combinations seen in a population inhabiting one locality seem to remain constant from year to year, although there are variations depending on the environment. It has for example been shown (Richards & Waloff, 1954) that the frequency of occurrence of green colours declined after a dry summer in which plants partially shrivelled, whilst brown forms increased in frequency. The explanation may well be that fewer green individuals survive a dry summer since their brown surroundings provide them with no protection against sighted predators. In general, however, it has to be said that the colours of field grasshoppers are very well adapted to their environment.

Fig. 39. Structure and terminology of the song of a gomphocerine grasshopper, *Chorthippus biguttulus* (L.). — The scale line below is 15 sec for sequence of 2nd order, 2.9 sec for sequence of 1st order, 72.5 msec for chirp, 9.1 msec for syllable, and 0.9 msec for impulse. (After Elsner, 1974).

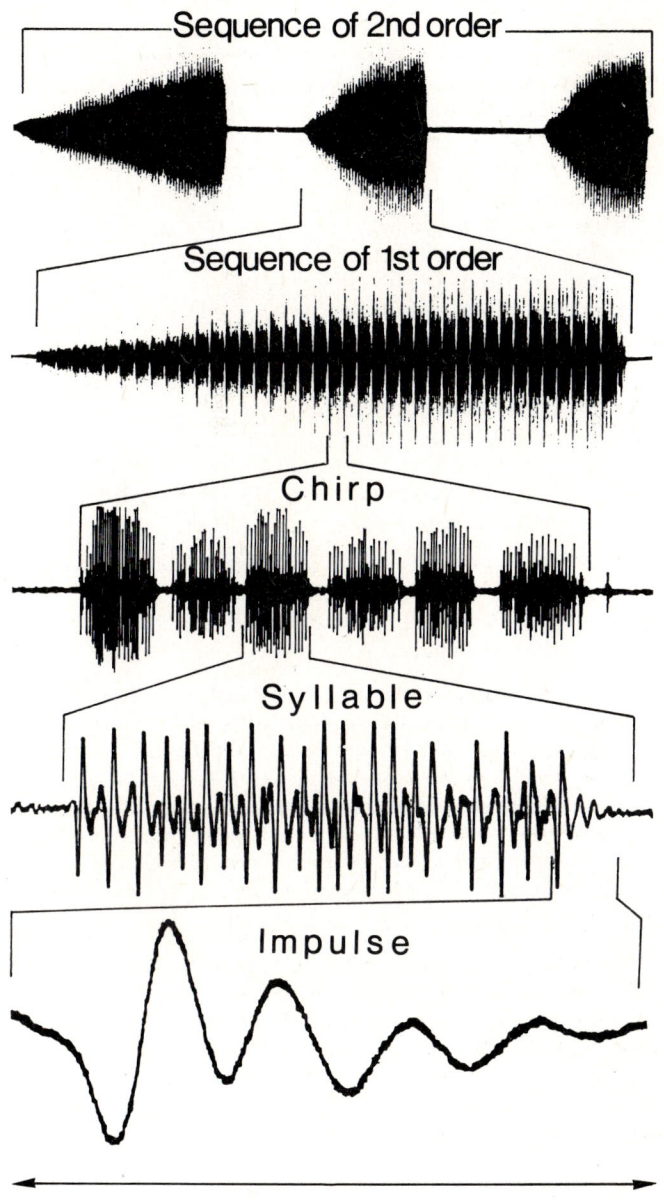

Sequence of 2nd order

Sequence of 1st order

Chirp

Syllable

Impulse

Stridulation

Stridulation in most species in Northern Europe is quiet, more or less merging into a rapidly reiterated hissing sound. The frequency of most recorded species ranges between 4,000 c/s (Hertz) and 40,000 c/s.

In many Gomphocerinae the male rubs either the left or right hind leg against the radius of the fore wing, or the legs are used alternately. The leg not being used in the stridulatory activity is moved as well.

The "calling song" is used as a long-distance signal in order to attract a co-specific female. The "courtship-song" is emitted when contact to a female is established, and has the aim of bringing her into a physiological condition for mating.

An oscillogram of sound production by the male in the "calling song" is reproduced. The following terminology (Elsner, 1974) is used.

1. Impulse. The emission resulting from the impact of one stridulatory peg at the vena radialis media of the tegmen. Such an impact elicits heavily damped oscillation. In acridids, mostly, the impact rate ("tooth-frequency") is smaller than the fundamental natural frequency of the stridulatory apparatus. However, the velocity of stridulatory movements and, thus, the impact rate varies considerably. It may reach the basic resonance frequency at certain times.

2. Syllable. The train of impulses produced by one femoral upward or downward movement. It has to be noted that this definition is more restricted than the one given by Jacobs and Broughton who define a syllable as the sound produced by one complete cycle of stridulatory movements.

3. Chirp. The series of syllables emitted by at least one complete cycle of upward and downward movements, i.e. the sound produced between the movement when the femur leaves the position of rest and the movement when it takes it up again. In the species described in the present paper, "a chirp is the shortest unitary rhythm element that can readily be distinguished as such by unaided human ear" (Broughton, 1963). In a few cases chirps are composed of just two syllables, one being produced by the upward and the other one by the downward movement. In most cases the intrachirp pattern is more complex: upstroke and downstroke are performed as vibrating or step-like movements resulting in a multisyllable chirp structure.

4. Sequence. A large number of chirps is linked together continuously (i.e. pauses are less than 100 msec), to form a 1st order sequence. In some species (e.g. *Chorthippus biguttulus)* a few of these units compose a 2nd order sequence.

Some grasshoppers can be difficult to identify. This applies to e.g. *Chorthippus brunneus, Ch. mollis* and *Ch. biguttulus.* It has even been doubted whether they belong to different species. The oscillograms do, however, demonstrate that they represent three quite separate species.

1 Pronotum extended a long way backward, out over the abdomen. Small insects.................................. Tetrigoidea (p. 67)
– Development of pronotum normal. Medium sized to large
   insects .... Acridoidea (p. 73)

Superfamily Tetrigoidea

(Ground-hoppers)

This group has been separated out as a superfamily of its own as a result of their divergent structure. Cf. the description under the family.
   Twelve species in four genera are known in Europe.

Family Tetrigidae

Small insects (about 10 mm). Pronotum extended posteriorly outwards over the abdomen. This extension covers the entire abdomen, reaching beyond its apex in all Northern European species. There is a more or less prominent median keel. The pronotum
and prosternum extend forwards, the head thus being surrounded by a collar. The fore
wings consist of small flaps, their normal function of protecting the hind wings having
been taken over by the extended pronotum. The hind wings are developed to a greater
or lesser extent (several non-Northern European species have reduced wings). No tympanal or stridulatory organs are found, even though ground-hoppers occasionally
move their hind legs up and down as if they were stridulating. The four valves of the
ovipositor are long and denticulate. The subgenital plate in the male is long and boat-
shaped.
   All Northern European species belong to one genus.

Genus *Tetrix* Latreille, 1802

Tetrix Latreille, 1802, Hist. Nat. Crust. Ins. 3: 284.
   Type-species: *Gryllus subulatus* Linnaeus, 1758.
Tettix Dalman, 1823, Anal. Entom., 86.

Small, powerfully built, short limbed animals. Integument granular and rough.
Paranota have two flaps at their hind edge. More or less monochrome animals, whose
colour may vary widely:- black, grey, brown, reddish-brown, yellowish or greenish.
   The number of larval instars in the male is five and in the female six. The first two

and the last three larval instars are homologous in the two sexes, the third larval instar in the female having no homologue in the male. Wing buds can be seen as early as the first larval instar, where they are situated as small lobes on the meso- and metanota along the sides of the thorax. In the following two instars in the male and three in the female the wing buds grow downwards. Not until the 4th and 5th instars in the male and the 5th and 6th in the female do the wing buds fold backwards and grow out over the abdomen. In order to see the wing buds it is necessary to cut the extended prono- tum away, since they lie concealed beneath it, except for the lowest portion of the rear wings in the final larval instars. Nymphs can be identified by the fact that the paranota have only one rearward flap and that the fore wings are concealed by the extended pronotum. (A key to the various instars is published by Farrow, 1964).

The appearance of the eggs differs from those of the grasshoppers by their having a horn at the top. They are laid in batches of from ten to twenty held together by a secretion, without it, however, forming a capsule or 'pod' as in the Acridoidea.

Ground-hoppers live close to the surface of the soil, generally preferring damp places. They are especially fond of low plants (moss, grass etc.) or the bare ground (e.g. sand). The nymphs inhabit the same places as the imagines. The diet usually consists of fine vegetable material (moss, algae) but may also include grass. They usually prefer sunny spots and are most active in daytime.

Their development diverges quite significantly from that of the grasshoppers, since they do not pass the winter as eggs. The eggs are laid during the summer in small depressions in the soil, in clumps of moss, cracks in bark etc. They develop into nymphs in the course of three to four weeks, the final larval stages passing the winter in clumps of moss or of grass, or under stones. The eggs which hatch may evolve into imagines in the course of the winter, or, more rarely, they may manage to develop to imago stage in the same year (e.g. *T. undulata*). They do not reach sexual maturity until the following summer, and they may be observed as early as the end of April, though not usually until May.

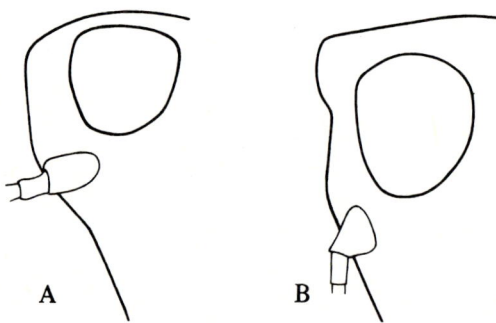

Fig. 40. Frons in profile of A: *Tetrix subulata* (L.) and B: *T. fuliginosa* (Zett.).

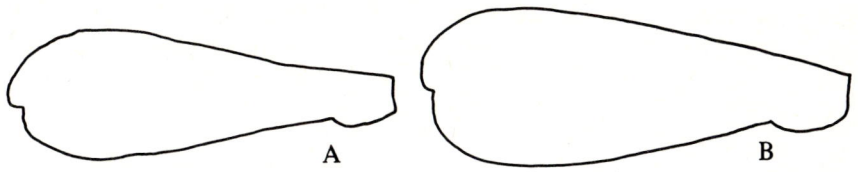

Fig. 41. Hind femur of A: *Tetrix undulata* (Sow.) and B: *T. bipunctata* (L.).

Key to species of *Tetrix*

1	Dorsal portion of pronotum almost level, with a weakly developed median keel 2
–	Dorsal portion of pronotum partially roof-like, with a prominent median keel, especially in the basal section 3
2 (1)	Frontal ridge seen in profile more or less flat between the eyes (Fig. 40A). Fastigium verticis 1^1/$_2$-2 times as wide as eyes seen from above . . . . 16. *subulata* (Linnaeus)
–	Frontal ridge seen in profile arched between the eyes (Fig. 40B). Fastigium verticis ^{1}/$_4$-^{1}/$_3$ the width of the eyes seen from above . . . . 17. *fuliginosa* (Zetterstedt)
3 (1)	Relatively slimly built. Median keel on pronotum not usually as highly curved as in next two species. Hind femur thin; 3-3^1/$_2$ times as long as it is wide (Fig. 41A). . . . . . 18. *undulata* (Sowerby)
–	Powerfully built. Median keel on pronotum usually highly curved. Hind femur broad; less than thrice as long as it is wide (Fig. 41B) 4
4 (3)	Longest segment of antenna twice as long as it is wide. Anterior border of pronotum obtuse-angled (Fig. 42A) 19. *bipunctata* (Linnaeus)
–	Longest segment of antenna three to four times as long as it is wide. Anterior border of pronotum practically straight (Fig. 42B) . . . . 20. *nutans* (Hagenbach)

Fig. 42. Head and pronotum in dorsal view of A: *Tetrix bipunctata* (L.) and B: *T. nutans* (Hag.).

16. *Tetrix subulata* (Linnaeus, 1758)
Figs 40A, 43.

Gryllus subulatus Linnaeus, 1758, Syst. Nat. ed. 10: 428.
Eng.: Slender Ground-hopper; Dan.: Sumptorngræshoppen; Fin.: Rantaokasirkka;
Sw.: Ängstorngräshoppan.

Colour may vary widely. Pronotum extends beyond apex of abdomen and out to over the middle of the hind tibia. Dorsal portion practically straight; median keel not particularly highly developed. Hind wings well developed, reaching at least to apex of pronotum. Distance between eyes 1½-2 times as wide as the eye, seen from above. Frontal ridge vertical; cannot be seen from above. Length to apex of pronotum: ♂ 9-12 mm; ♀ 11-15 mm.

Distribution. The widest distributed species of *Tetrix* within the area concerned. The only places it does not seem to occur are Swedish and Finnish Lapland and certain sites on the West coast of Norway. — The total distribution comprises almost the whole of Europe.

Biology. Usually prefers damp places; often found near lakes and rivers. Good flier, frequently taking to the air to escape (Ander, 1947).

17. *Tetrix fuliginosa* (Zetterstedt, 1828)
Fig. 40B.

Acridium fuliginosum Zetterstedt, 1828, Ins. Lapp. 1: 452.
Fin.: Lapinokasirkka; Sw.: Nord-torngräshoppan.

Colour may vary widely. Pronotum extends beyond apex of abdomen and out to over the middle of the hind tibia. Dorsal portion practically straight; median keel not par-

Fig. 43. *Tetrix subulata* (L.) ♀.
Length 11-15 mm.

ticularly highly developed. Hind wings well developed, reaching at least to apex of pronotum. Distance between eyes $1/4$-$1/3$ of width of eye, seen from above. Frontal ridge concave between the eyes, seen from the side. Length to apex of pronotum: ♂ ♀ 11.5-17 mm.

Distribution. Holarctic. In the northern parts of Sweden and Finland; in Sweden, S. to Dlr. and Hls., in Finland, S. to Ok and ObS. Probably also to be found in N. Norway. The distribution extends eastwards over the northern parts of the USSR to Sakhalin.

Biology. Damp places with meadow vegetation and bushes (Lindberg, 1952; Valle, 1930).

18. *Tetrix undulata* (Sowerby, 1806)
    Figs 41A, 44.

Acridium undulatum Sowerby, 1806, Brit. Misc. 2: 28.
Eng.: Common Ground-hopper; Dan.: Den almindelige torngræshoppe; Fin.: Ketookasirkka; Sw.: Glänt-torngräshoppan.

Quite robustly built, although slimmer than the next two species. Colours vary widely. Pronotum upwardly vaulted, with a sturdy median keel the upper edge of which, seen from the side, is reasonably straight. Pronotum does not normally extend as far as the hind tibia. Hind wings about twice as long as fore wings, not extending to apex of pronotum. Hind femur narrow (three times as long as it is wide). Length to apex of pronotum: ♂ ♀ 8-15 mm.

Fig. 44. *Tetrix undulata* (Sow.) ♀.
Length 8-15 mm.

71

Distribution. Commonly distributed throughout the southerly regions of the area, reaching as far north as Central Sweden, southwestern Finland and southern Norway. — The commonest species of *Tetrix* in northwestern Europe including The British Isles, France, N. Spain, C. Europe, Rumania and western USSR.

Biology. Frequently found in large numbers at the edges of and in clearings in woods between the leaves, in the grass and in clumps of moss. In meadows and bogs, in addition to quite dry places.

19. *Tetrix bipunctata* (Linnaeus, 1758)
Figs 41B, 42A.

Gryllus bipunctatus Linnaeus, 1758, Syst. Nat. ed. 10: 427.
Eng.: Two-spotted Groundhopper; Dan.: Den toplettede torngræshoppe; Fin.: Nummiokasirkka; Sw.: Tvåpunkterad-torngräshoppan.

Robustly and powerfully built. The colours may vary widely, the name referring to two quite frequently prominent black spots on the upper surface of the pronotum, though they are not found in all specimens. The pronotum is upwardly vaulted, with a sturdy median keel, the upper edge of which, seen from the side, is sharply curved upwards. Pronotum does not normally extend as far as to the base of the hind tibia. Anterior edge of pronotum is angular when seen from above. Hind femur wide (under three times as long as it is wide). Antennae approximately equal in length to fore femora; their middle segments twice as long as they are wide. Length to apex of pronotum: ♂ ♀ 8-11 mm.

Distribution. Rare in Denmark: only recorded from SJ (Vojens), NWJ (Sevel) and Bornholm (Almindingen). — Generally distributed in Sweden, Norway, Finland and adjacent areas of the USSR. — France, C.Europe, Rumania, USSR, Mongolia, NE.China.

Biology. This species lives in dry, warm sites at open spaces in woods, on moors — including in mountainous country — and slopes.

20. *Tetrix nutans* (Hagenbach, 1822)
Fig. 42B.

Tetrix nutans Hagenbach, 1822, Symb. Faun. Ins. Helvet.: 41.
Fin.: Tarhaokasirkka.

Robustly and powerfully built. Colour may vary widely. Pronotum upwardly vaulted with a sturdily developed median keel, the upper edge of which, seen from the side, is sharply curved upwards. Pronotum does not normally extend as far back as to the base of the hind tibia. Anterior edge of pronotum, seen from above, is rounded to nearly straight. Hind femur broad (less than 3 times as long as it is wide). Antennae 1½ times as long as fore femora, their middle segments being 3-4 times as long as they are wide. Length to apex of pronotum: ♂ ♀ 9.5-10 mm.

This species is divided up into two subspecies:- *nutans* s.str., in which the middle segments of the antenna are at least 4 times as long as they are wide, and *tenuicornis* (Sahlberg, 1893), in which they are only about 3 times as long as wide. The populations found in Finland belong to the subspecies *tenuicornis*.

Distribution. Not found in Denmark, Sweden or Norway. — Scattered distribution in southeastern Finland and the adjacent areas of the Soviet Union. One single individual has been found near Mölln, in southern Holstein. — France, C. Europe, USSR, Mongolia and China.

Biology. Prefers sites offering sparse vegetation, e.g. clearings, cliffs, and marshy places.

Superfamily Acridoidea

(Grasshoppers)

Body more or less cylindrical in structure, commonly compressed from the sides. Several groups possess a highly developed stridulatory organ. Tympanal organ mounted on sides of first abdominal segment. In other respects the species in this superfamily are built as described in the introductory paragraphs to the Caelifera.

Four families known in Europe, only two of which have representatives in Northern Europe.

Key to families and subfamilies of Acridoidea

1   Fore wings reduced to small, elliptical flaps. No lateral keels
     on pronotum. Prosternum bearing a conical process between
     the coxae. . . . . Catantopidae/Catantopinae (p. 74)
–   Fore wings normal; if they are short, there are prominent
     side keels on the pronotum. No conical process on proster-
     num. . . . . Acrididae (p. 76) 2
2 (1) Vena intercalata in median area of fore wing present — at
     least in the male — and bearing the file in many species
     (Fig. 38). Frons and vertex normally at a right angle. Hind
     wing usually highly coloured. *(Mecostethus* is the only excep-
     tion in N.Europe: it can be identified by the fact that the
     foveolae are triangular and vena intercalata is present) . . . . Locustinae (p. 76)
–   Vena intercalata absent. File situated on inner side of hind
     femur (Fig. 37). Frons and vertex at an acute angle. Hind
     wing transparent. Foveolae, if present, rectangular. . . . Gomphocerinae (p. 85)

Family Catantopidae

This family is characterised by a highly variable construction of the body, the wings, the cerci in the male, and the epiproct. Subdivided into a number of subfamilies. We shall in this publication only treat the two species occurring in N. Europe. They both belong to the Catantopinae.

SUBFAMILY CATANTOPINAE

Medium-sized insects. Frons normally vertical. No side keels on pronotum in species found in Scandinavia. No specialised stridulatory organ. Tympanal organ present. Subgenital plate in male long and pointed.

Numerous genera and species throughout the world. In Europe, 14 genera, comprising 52 species.

Key to genera of Catantopinae

1   Prozona same length as metazona, or only slightly longer.
     Hind margin of pronotum straight or only slightly rounded.
     Hind tibia bluish, with white spines . . . . *Podisma* Berthold (p. 75)
−   Prozona significantly longer than metazona. Hind margin
     of pronotum obtuse angled. Hind tibia reddish . . . . . . *Melanoplus* Stål (p. 75)

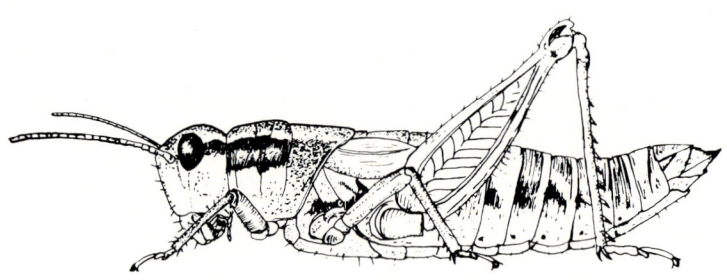

Fig. 45. *Podisma pedestris* (L.) ♀.
Length 24-30 mm. (From Harz, 1975).

Genus *Podisma* Berthold, 1827

Podisma Berthold, 1827, *in* Latreille's Fam. Thierr.: 441.
Type-species: *Gryllus pedestris* Linnaeus, 1758.

Frontal costa only sunken around the ocellus. Fastigium verticis about 1.5 times as wide as an eye seen from above. Hind margin of pronotum straight or slightly convex. Prozona same length as metazona or only slightly longer. Prozona smooth, with two transverse furrows. Metazona coarsely punctate. Tympanum uncovered.

21. *Podisma pedestris* (Linnaeus, 1758)
Fig. 45.

Gryllus pedestris Linnaeus, 1758, Syst. Nat. ed. 10: 433.
Eng.: Forest Grasshopper; Dan.: Skovgræshoppen; Fin.: Kangassirkka; Sw.: Skogs-gräshoppan.

Basic coloration, reddish brown, yellowish to blackish. Paranota dorsally bearing a longitudinal dark band, under which there is a yellow spot. Hind tibia greyish blue to blue, with white spines. Length: ♂ 17-19 mm; ♀ 24-30 mm.

Distribution. Not found in Denmark. — Generally distributed, though uncommon, in Sweden, Norway and Finland. — A Euro-Siberian species, distributed from N.Scandinavia southward to the mountains of Italy, France and Greece; westward to the Pyrenees.

Biology. May be found on e.g. grass or heather; especially prefers sunny slopes near cliffs and stones, and slightly gravelly ground. Both males and females are capable of producing a clicking sound by rubbing the mandibles against one another.

Genus *Melanoplus* Stål, 1873

Melanoplus Stål, 1873: 79.
Type-species: *Acrydium femur-rubrum* De Geer, 1773.

Frontal costa with a deep furrow round the ocellus. Width of fastigium verticis almost the same as that of the eye, seen from above. Hind margin of pronotum convex. Prozona significantly longer than metazona. Prozona smooth, with two transverse furrows. Metazona coarsely punctate. Normally brachypterous. Fore wings lanceolate. Tympanum uncovered.

22. *Melanoplus frigidus* (Boheman, 1846)

Gryllus frigidus Boheman, 1846, Öfvers. K. Sv. Vet.Akad.Förh.: 80.
Eng.: Mountain Grasshopper; Dan.: Fjeldgræshoppen; Fin.: Lapinsirkka; Norw.: Fjellgrashoppe; Sw.: Fjällgräshoppan.

Basic colour brownish to black. Paranota dorsally with a more or less well marked, elongated yellow spot. Inner surfaces of hind femora red. Hind tibia red with black spines. Length: ♂ 17-20 mm; ♀ 24-26 mm.

Distribution. Not found in Denmark. — In Scandinavia, associated with mountainous regions, being found from the south of Norway to the north. — In Sweden, from Härjedalen and Jämtland to Lapland. — In Finland, it is found in Karelia and the corresponding region of Russia, as well as in the northerly provinces and on the Kola Peninsula.

The species seems to have survived the last Ice Age on the tundra below the ice cap. When the ice retreated, it invaded the Scandinavian mountain chain. It is also found in the Alpine regions of Central Europe and in isolated sites in the northern parts of the Soviet Union and northern Canada (Ander, 1947).

Biology. Inhabits both Arctic and Subarctic regions of the mountain chain. In Finland, it has also been found in the birch and pine forest region. It lives on mountain moorlands and low willow bushes. Eggs laid in the summer commence development the same autumn, hatching the moment the weather warms up the following summer.

Family Acrididae

Head variable in form, but of normal construction in Northern European species. Prosternum lacks the conical process between the coxae. Stridulatory organ usually present.

For key to the subfamilies, see p. 73.

SUBFAMILY LOCUSTINAE

From quite large to large insects (c. 15-70 mm). Fore wings more or less brown spotted; hind wings frequently highly coloured. If foveolae present, they are triangular or pentagonal. Frons and vertex at right angles to each other, though not in *Mecostethus*. Integument usually rough (rugose) and knobbly. In species in which a stridulatory organ has evolved, the file is sited on the vena intercalata and scrapes against the inner edge of the hind femur.

50 species have been found in Europe, belonging to 21 genera.

Key to genera of Locustinae

1   Frons and vertex at an acute angle to each other . *Mecostethus* Fieber (p. 83)
–   Frons and vertex at a right angle to each other 2
2 (1) Only a suggestion of a median keel on pronotum in the me-
        tazona . . . . *Sphingonotus* Fieber (p. 83)

Genus *Psophus* Fieber, 1853

Psophus Fieber, 1853, Lotos 3: 122.
   Type-species: *Gryllus stridulus* Linnaeus, 1758.

No foveolae. Median keel of pronotum upwardly vaulted and not intersected by transverse sulci. The transverse sulcus present ends in a depression in the middle of the pronotum close to the median keel. Upper edge of hind femur straight and non-denticulate. Wings well developed, extending beyond posterior tip of abdomen in male and not quite so far in female.
   Only one species in Europe.

23.   *Psophus stridulus* (Linnaeus, 1758)

Gryllus stridulus Linnaeus, 1758, Syst. Nat. ed. 10: 432.
Dan.: Den trommende græshoppe; Fin.: Palosirkka; Norw.: Klapregrashoppe; Sw.: Trumgräshoppan.

Coloration may vary between black, grey-brown and red to yellowish brown. Lighter-coloured spots on fore wings. Hind wings pink, with black to blackish brown tips. Length: ♂ 23-25 mm; ♀ 26-40 mm.

   Distribution. Not found in Denmark. — In Sweden, from Sk. north to Dlr. and Gstr.; in Norway, in some southeastern districts. — Finland: from Ab to Om and Ok, but not Al. Also in adjacent areas of the USSR. Everywhere sporadic and uncommon. — The range is from the Pyrenees and the higher parts of France, eastward through the forest and wooded steppe zones of Europe and Asia.

   Biology. This species particularly demands high temperatures, inhabiting sunny spots such as sandy plains, clearings in forests, and heathland.

The male is a good flier and can cover long distances, making a drumming noise with his hind wings. The female is clumsier but can jump up to about a metre with the aid of her wings. They are easy to identify in flight by the prominent red coloration of the hind wings.

Genus *Locusta* Linnaeus, 1758

Locusta Linnaeus, 1758, Syst. Nat. ed. 10: 431.
Type-species: *Gryllus migratorius* Linnaeus, 1758.

Foveolae triangular, flat, rather inconspicuous. Median keel on pronotum more or less projecting and intersected by a transverse sulcus. Wings well developed, extending beyond the hind femora. Hind wings transparent, slightly yellowish in colour and with no prominent coloration or dark banding (tips slightly smoky in colour). Upper edge of hind femur straight and slightly denticulate.

24. *Locusta migratoria* (Linnaeus, 1758)

Gryllus migratorius Linnaeus, 1758, Syst. Nat. ed. 10: 432.
Gryllus danicus Linnaeus, 1767, Syst. Nat. ed. 12: 702.
Eng.: Migratory Locust; Dan.: Den almindelige vandregræshoppe; Norw.: Europeisk vandregrashoppe; Sw.: Europeiska vandringsgräshoppan.

This species may occur in three phases, of which one *(transiens)* is transient and the other two are main phases *(solitaria* and *gregaria).* These two forms are so different that they were at one time regarded as separate species.

Key to phases of *L. migratoria*

1 Median keel on pronotum projecting and upwardly vaulted. Pronotum narrows anteriorly, the front edge being pointed. . . . . 'Phase *solitaria*'
– Median keel on pronotum straight or slightly down-curved. Pronotum only slightly narrowed anteriorly, the front edge being straight . . . . 'Phase *gregaria*'

A. 'Phase *solitaria*', the inconspicuous, non-migratory phase. Overall usually green or brown in colour. Pronotum comparatively long, narrowed anteriorly, the prozona being longer than the metazona. In lateral view, the median keel projects and is upwardly vaulted, the front and edges of the pronotum running angularly out into a series of points. Vertex flat, with median keel either absent or only weakly developed. Fore wings slightly shorter than in the plague phase and covered with numerous tiny brown spots. Colour of hind tibia varies. The two sexes are of very different sizes (♂ 29-37 mm; ♀ 35-51 mm).

78

B. 'Phase *gregaria'*, the active, excitable, gregarious plague phase. Usually bright yellow, grey and brown in colour. Pronotum comparatively short; only slightly narrowed anteriorly. Prozona shorter than metazona. Median keel straight or slightly downcurved in lateral view. Front edge of pronotum straight, the hind edge not forming as acute an angle as in the non-plague phase. Vertex upwardly vaulted, with a median keel. Fore wings slightly longer than in the 'phase *solitaria*', with brown spots. Hind tibia yellow. Hind legs not quite so powerfully developed as in the 'phase *solitaria*'. The two sexes approximately equally sized (♂ 35-50 mm; ♀ 42-55 mm).

Distribution and biology. This is the most widely distributed of all locust species. It is found in Africa, Europe, Asia and northern Australia. Northward, it has been observed almost up to the Arctic Circle.

The common migratory locust has evolved into a number of races, the one occurring in Northern Europe belonging to *Locusta migratoria migratoria*. Its natural habitat is the river valleys and deltas of the Black Sea, Caspian Sea and Sea of Aral region, where plagues used to occur. I. e. plagues used to take place in Central Europe in former times, the non-plague 'phase *solitaria*' being capable of reproducing and surviving in certain localities (southern Holland being the northern limit).

The first description of *L. migratoria* 'phase *solitaria*' was published by Linnaeus in 1767, using an individual captured in Denmark *(Gryllus danicus)*. It migrates to Fennoscandia and Denmark from time to time, both as the 'phase *solitaria*' and (more frequently) as the 'phase *gregaria*'. A huge swarm was for example described in southern Skåne in the summer of 1846, spreading to the Gothenburg area, and the following year it occurred in Skåne again. It is unable to survive as the 'phase *solitaria*' in Fennoscandia and Denmark.

Genus *Oedipoda* Serville, 1831

Oedipoda Serville, 1831, Ann. Sci. Nat. 22: 287.
Type-species: *Gryllus caerulescens* Linnaeus, 1758.

Foveolae irregularly triangular or pentagonal. Median keel on pronotum conspicuously divided by a transverse sulcus. Well developed wings extending beyond apex of hind femora. Hind wings usually brightly coloured red, blue or yellow, with a black band at the edge. Upper edge of hind femur nondenticulated, and sharply curved downward a little distal to the middle.

25. *Oedipoda caerulescens* (Linnaeus, 1758)
Fig. 46.

Gryllus caerulescens Linnaeus, 1758; Syst. Nat. ed. 10: 432.
Dan.: Den blåvingede ørkengræshoppe; Fin.: Kirjosiipisirkka; Sw.: Blåvingade sandgräshoppan.

Hind wings blue with a black band at the edge, with a wedge-shaped prolongation towards the base of the wing near its anterior edge. Median keel slightly higher in prozona than in metazona. Considerable variation in colour in this species from quite light to black, depending on the sort of soil it inhabits. It usually remains unobserved until it takes flight and flies up, showing its highly coloured blue hind wings. Length: ♂ 15-21 mm; ♀ 22-28 mm.

Distribution. A certain number of finds were published in the 19th Century in Denmark, from Sønderborg in South Jutland, and in Sweden, on the island of Särö in Halland. In 1974 a female was captured near Boderne on the island of Bornholm. This was presumably a new migrant (Johnsen, 1976). Found at several sites in Holstein round Lübeck, Ratzeburg and Mölln. — Central and Southern Europe, USSR, North Africa, Asia Minor, Central Asia to China.

Biology. On Bornholm, it was found on sandy soil near a heathery dune — a characteristic habitat for this species. It lives in dry regions in sites more or less without vegetation, frequently on stony or sandy soil.

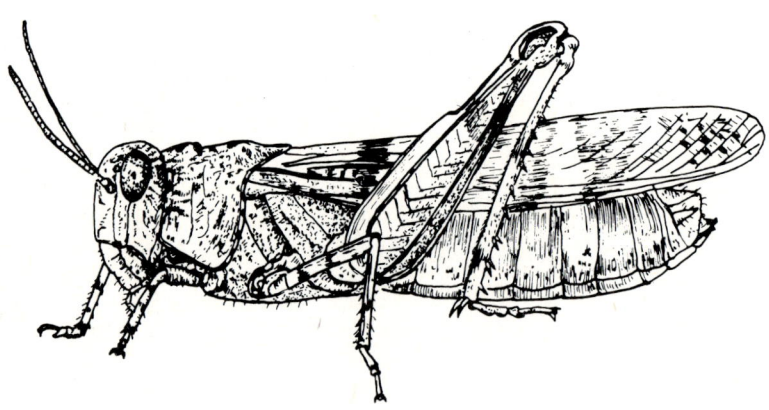

Fig. 46. *Oedipoda caerulescens* (L.) ♀.
Length 22-28 mm. (From Harz, 1975).

Genus *Bryodema* Fieber, 1853

Bryodema Fieber, 1853, Lotos 3: 129.
  Type-species: *Oedipoda gebleri* Fischer-Waldheim, 1836.

Foveolae large, with indistinct margins. Wings well developed, extending beyond apex of hind femora. Fore wings have an incomplete vena intercalata in the medial area. Median keel on pronotum weakly developed and interrupted by two transverse sulci. Metazona much larger than prozona. Upper edge of hind femur non-denticulate and straight.

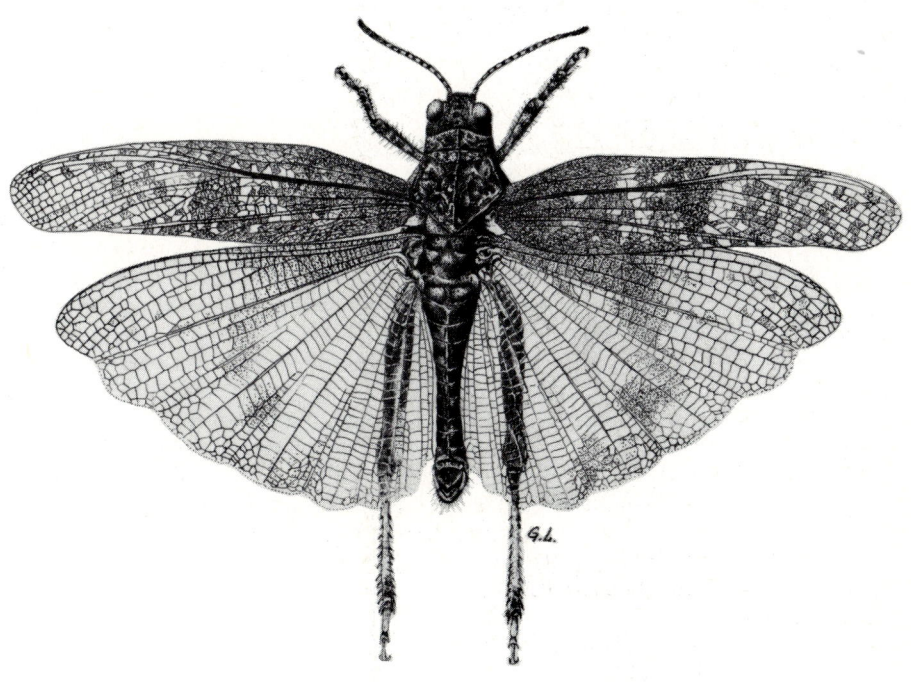

Fig. 47. *Bryodema tuberculata* (F.) ♂.
Length 27-31 mm.

26. *Bryodema tuberculata* (Fabricius, 1775)
Fig. 47.

Gryllus tuberculatus Fabricius, 1775, Syst. Ent.: 290.
Dan.: Hedeskratten; Fin.: Ruususiipisirkka; Sw.: Rosenvingade gräshoppan.

Head and pronotum rugose and punctuate. Basic coloration greyish brown to yellowish, with varying brightly coloured parts. Some of the lighter coloured areas may even be green. Inner area of hind wings bright red in colour (particularly the transverse veins), with an outer border of dark brown. Tips of hind wings transparent, with brown veins. Hind femora yellowish brown, with two dark spots on upper side. Hind tibiae yellow. Length: ♂ 27-31 mm; ♀ 33-36 mm.

The species is divided into a number of subspecies, the European representative belonging to *tuberculata* s. str.

Distribution. In this area, present only in Sweden, on Öland, where it is common on the "alvar" heath. It used to be common in Denmark on the moorlands of Central and West Jutland, but their reclamation for agriculture has reduced it enormously, and it may completely have disappeared from the fauna of Denmark. The last finds were made on Abild Hede, near Varde Å, in 1949; near Silkeborg in 1941, and on Tandrup Hede near Løgstør, also in 1941. — A typical Continental insect, with its main area of distribution in the Soviet Union and the temperate zone of Asia. Only the highly specialised ecology of the "alvar" of Öland makes it possible for this species to survive there, in the same way that the moorlands of Jutland — produced as they were by the hand of Man — provided it with a basis for existence.

Biology. This species has long been known to be an inhabitant of the sandy soils of Jutland. It is found in level moorlands with a stunted flora consisting of the shorter Ericaceae, slender *Sarothammus*-bushes, *Arctostaphylos, Vaccinium,* various Compositae and grasses. Between and under these plants are patches of bare soil, clumps of moss and lichen (Nørgaard, 1942). This description applies to Tandrup Hede, near Løgstør, where the species was very common over an area of a few hundred square metres. Its coloration camouflages it excellently against the heathland vegetation against which it flattens itself, but if one approaches too closely it flies up with a rattling metallic sound, the red hind wings becoming visible at the same time. It can fly for quite long distances (over 40 m), but usually flies less. The altitude of flight is usually 1-1.6 m, but it has been known to get up to over 5 m. While in flight it alternates between active flight and gliding. When the wings are in motion, the rattling sound is produced by the downward beat of the hind wings, perhaps even beating against each other. It is usually the male which flies up, whilst the female flies more rarely and only over short distances. It makes the rattling sound only at the commencement of flight, and is less noisy than the male.

The eggs are laid in the soil.

Genus *Sphingonotus* Fieber, 1852

Sphingonotus Fieber, 1852, *in* Kelch: Orth. Oberschlesiens: 2.
   Type-species: *Gryllus caerulans* Linnaeus, 1767.

Foveolae triangular and elongate. Wings well developed, extending beyond apex of hind femora. Anterior portion of pronotum greatly constricted, with three transverse sulci. Median keel only present on metazona and only slightly projecting.

27. *Sphingonotus c. caerulans* (Linnaeus, 1767)

Gryllus caerulans Linnaeus, 1767, Syst. Nat. ed. 12: 701.
Dan.: Den blåvingede steppegræshoppe; Fin.: Sinisiipisirkka; Sw.: Blåvingade gräshoppan.

Hind wings transparent and uniformly blue in colour. Inner edge of hind femur has a light-coloured stripe.

   This species is divided into a number of subspecies, Scandinavian specimens belonging to *cyanopterus* (Charpentier, 1825), which have a mere suggestion of a transverse band or stripe on the hind wings. Length: ♂ 15-19 mm; ♀ 20-25 mm.

   Distribution. Not found in Denmark.— Sweden: inhabits the "alvar" areas of Öland and Gotland, and also occurs sporadically in the coastal region from Gothenburg in W. Sweden and on up into Norway, where its southwesterly extension reaches Aust-Agder. — Finland: Ab and N. — In Holstein it has been found at various localities, e.g. south of Mölln. — Central Europe, N. France, NW. USSR.

   Biology. Inhabits dry places in dunes, on steppes and in "alvar" areas where the vegetation is very sparse. Everywhere very localised. It is highly probably a steppe relict in N. Europe.

   The species is common in localities near the shore on Hangö in S. Finland as well as on four neighbouring islands which are now part of the Soviet Union. It seems to have been distributed by ice or water in the egg stage. Males can fly some 1-2 m at a height of some 20 cms., and females even less (Nordman, 1963).

Genus *Mecostethus* Fieber, 1852

Mecostethus Fieber, 1852, *in* Kelch: Orth. Oberschlesiens: 1.
   Type-species: *Gryllus grossus* Linnaeus, 1758.
Stethophyma Fischer, 1853.

Foveolae triangular, often indistinct. Pronotum not constricted in the middle. Upper surface of pronotum bears a median keel and two practically straight side keels which are wider and less prominent than it. Wings extend beyond distal end of hind femora.

Vena intercalata in the medial area straight, prominent, and nearer the cubitus than the media. Subgenital plate conical in male. Valves of ovipositor long.

The general characteristics point in the direction of the Gomphocerinae, but the vena intercalata is prominent and bears those pegs which are regarded as a significant characteristic of the Locustinae.

28. *Mecostethus grossus* (Linnaeus, 1758)
Figs 35A, 48.

Gryllus grossus Linnaeus, 1758, Syst. Nat. ed. 10: 433.
Eng.: Large Marsh Grasshopper; Dan.: Sumpgræshoppen; Fin.: Suosirkka; Sw.: Kärrgräshoppan.

The basic colouration is olive-green which may, however, shade over into yellowish or brownish. The side keels on the pronotum are blackish brown. The venation of the fore wings is blackish brown between radius and cubitus, most strongly coloured towards the base. There is a yellow stripe along the leading edge of the fore wings. The hind wing are colorless, with a blackish brown net of veins in the anterior region. The hind femora are red underneath, and the hind tibiae are a sulphurous yellow with black spines. Length: ♂ 12-25 mm; ♀ 22-39 mm.

Distribution. Occurs sporadically in Denmark and Fennoscandia. Not found in the N. of Norway, some of the mountainous regions of S. Norway, nor in the northernmost parts of Finland or the neighbouring areas of Russia. — The British Isles, Central Europe, France, N.Spain, N.Italy and N.Yugoslavia; also in the mountains of Rumania and Bulgaria, and further into the USSR.

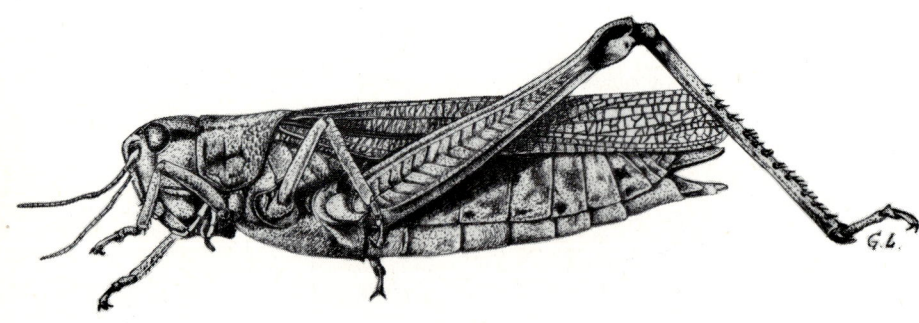

Fig. 48. *Mecostethus grossus* (L.) ♀.
Length 22-39 mm.

Biology. On damp grassy spots in meadows and by lakes and streams. Whether a bog is surrounded by woodland or open country seems to have no effect on the occurrence of this species. Flies willingly on hot days. The male has been observed to fly 20-30 m (Ander, 1947).

The egg capsules (pods) are 10-16 mm in length, their surface consisting of a hardened secretion. A pod may contain up to 40 eggs, which are arranged diagonal to the longitudinal axis of the pod. The contours of the eggs are visible through the surface. The eggs are 6.1-7 mm in length, reddish brown, and hexagonal in structure. The pods are laid in the soil or on clumps of grass (Richards & Waloff, 1954).

Stridulation. A sound is produced by the terminal spines on one — rarely both — of the hind femora striking the distal portion of the fore wing at rest. This results in a short, sharp "tick", which is usually repeated in series of two or three "ticks" a second and which are audible up to several metres' distance.

SUBFAMILY GOMPHOCERINAE

Medium-sized insects (c. 10-30 mm). Fore wings frequently either colourless or weakly colored, transparent, and with prominently visible veins. Hind wings colourless and transparent. Foveolae absent, or rectangular if present. Frons and vertex at an acute angle to each other. No downward pointing peg on prosternum between the fore coxae. The best identificatory feature of this subfamily is the stridulatory organ, in which the file is mounted on the inner of the hind femur and the scraper comprises the radius of the fore wing. No vena intercalata on the median area of the fore wings.

Key to genera of Gomphocerinae

1   Foveola absent (Fig. 52). . . . . *Chrysochraon* Fischer (p. 87)
–   Foveola present 2
2 (1) Antennae filiform (Fig. 49) 3
–   Antennae clublike in lateral view, markedly so in the male (Fig. 50) 6
3 (2) Anterior edge of fore wings straight (Fig. 51A) 4
–   Anterior edge of fore wings slightly widened towards the base (Fig. 51B) . . 5
4 (3) Valves of ovipositor each bearing a "tooth" (Fig. 54). Side
      keels regularly introflexed (greatest distance between them
      1¼-1½ times that at narrowest portion) . . . . . . *Stenobothrus* Fischer (p. 88)
–   Valves of ovipositor lack "teeth". Side keels angularly in-
      troflexed (greatest distance between them 2-3 times that at
      narrowest portion) . . . . *Omocestus* Bolivar (p. 90)
5 (3) Cubitus 1 and 2 of fore wings fused and medial area highly
      broadened in the male (Fig. 63) . . . . *Stauroderus* Bolivar (p. 95)

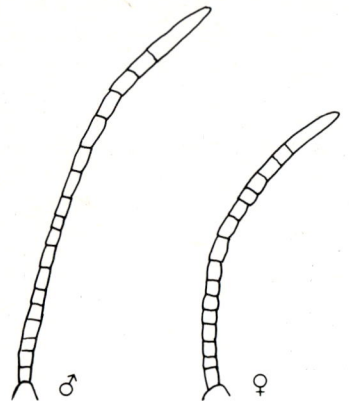

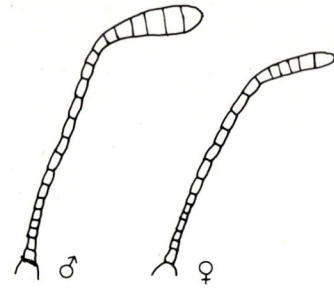

Fig. 50. Antennae of *Myrmeleotettix maculatus* (Thnbg.).

Fig. 49. Antennae of *Chorthippus brunneus* (Thnbg.).

–   Cubitus 1 and 2 of fore wings separate (although fused ex-
     cept for the initial portion after the origin in *Ch. apricarius)*
     and medial area either slightly developed or totally undevel-
     oped . . . . *Chorthippus* Fieber (p. 96)
6 (2) Leading edge of fore wings straight (Fig. 51A) *Myrmeleotettix* Bolivar (p. 113)
–   Leading edge of fore wings with a slight enlargement near
     the base (Fig. 51B) 7
7 (6) Distal club of antennae broad, with a white tip. Tympa-
     num approximately half obscured . . . . *Gomphocerus* Thunberg (p. 114)
–   Distal club of antennae narrower and uniform in colour.
     Tympanum approximately one third obscured . *Aeropedellus* Hebard (p. 116)

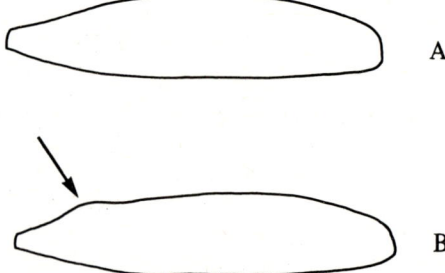

A

Fig. 51. Shape of fore wing of A: *Myrmeleotettix maculatus* (Thnbg.) and
B: *Gomphocerus rufus* (L.).

B

Genus *Chrysochraon* Fischer, 1853

Chrysochraon Fischer, 1853; Orth. Eur.: 296, 307.
  Type-species: *Podisma dispar* Germar, 1831-35.

Foveolae absent. Antennae filiform. Side keels on pronotum weakly developed and practically straight. The valves of the ovipositor short and non-denticulate. Subgenital plate in male pointed and conical in shape.

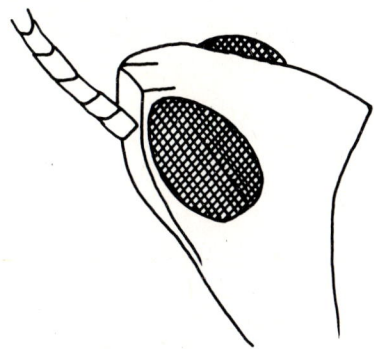

Fig. 52. Head of *Chrysochraon dispar* (Germ.).

27. *Chrysochraon d. dispar* (Germar, 1831-35)
  Fig. 52.

Podisma dispar Germar, 1831-35, Fauna Ins. Eur. fasc. 17, Taf. 7.
Dan.: Guldgræshoppen; Fin.: Kultaheinäsirkka; Sw.: Guldgräshoppan.

Fore wings in male wide, rounded, reaching approximately to the tip of the abdomen, whilst in the female they are shorter and lanceolate. Hind wings highly reduced. The male is a shiny yellowish green and the female brownish grey with a silky sheen. The hind tibia and the underneath of the hind femora are bright red in colour in the female. Length: ♂ 16-19 mm; ♀ 22-29 mm.

  Distribution. Not found in Denmark or Norway. — In Sweden, it occurs on Öland and Gotland, in Mälardalen (Vstm.) and in the neighbourhood of Luleå(Nb.). — In Finland, it has been found at several sites along the Gulf of Bothnia and in the south of Finland as well as in Karelia and the corresponding regions of Russia. — Found at certain localities in Holstein in the neighbourhood of Lübeck. — France, C.Europe, Yugoslavia, Hungary, Rumania and USSR.

  Biology. In damp places, with *Carex* and grassy vegetation. Pods of eggs laid in the stems.

Stridulation. The sequence lasts about a second. Each sequence consists of some 15 chirps, with a metallic sound (Wallin, 1979).

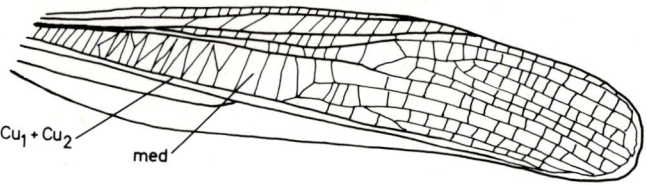

Fig. 53. Fore wing of *Stenobothrus lineatus* (Panz.).

Genus *Stenobothrus* Fischer, 1853

Stenobothrus Fischer, 1853; Orth. Eur.: 296, 313.
    Type-species: *Gryllus lineatus* Panzer, 1796.

Foveolae narrow and rectangular. Antennae filiform. Side keels on pronotum weakly angularly introflexed (greatest distance between them 1¼-1½ times greater than at the narrowest place). Leading edge of fore wing straight, with no widening at the base. The most significant identificatory feature of this genus is the fact that each of the valves of the ovipositor is equipped with a "tooth".

Key to species of *Stenobothrus*

1   Fore wings broad and short. Cubitus-1 and cubitus-2 fused . . . .
    . . . . 30. *lineatus* (Panzer)
–   Fore wings narrow, extending in the ♂ to the tip of the ab-
    domen or a little further. Cubitus-1 and cubitus-2 not
    fused . . . . 31. *stigmaticus* (Rambur)

Fig. 54. Ovipositor of *Stenobothrus lineatus* (Panz.).

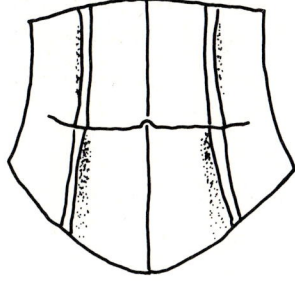

Fig. 55. Pronotum of *Stenobothrus lineatus* (Panz.).

30. *Stenobothrus lineatus* (Panzer, 1796)
Figs 53-55.

Gryllus lineatus Panzer, 1796, Faun. Ins. Germ. 33: 9.
Eng.: Stripe-winged Grasshopper; Fin.: Juovaheinäsirkka.

Coloration highly variable, from green to yellow, brown and red. Fore wings broad, only extending a short way beyond the tip of the abdomen. Medial area (med) of fore wings greatly widened; cubitus-1 and cubitus-2 fused. Length: ♂ 16-18 mm; ♀ 21-24 mm.

Distribution. Not found in Fennoscandia or Denmark. — Nearest occurrence to the area: Holstein, where it is found in several places (Segeberger Heide near Neumünster and in a southern area from Ratzeburg over Mölln to Lauenburg and Hamburg). — Southern England, France, Central Europe, Hungary, Rumania, Bulgaria, mountainous regions of the Mediterranean countries, USSR and Mongolia.

Biology. In dry places, e.g. in woodland glades and on heathland.

31. *Stenobothrus stigmaticus* (Rambur, 1838)

Gryllus stigmaticus Rambur, 1838, Faune Andal. 2: 93.
Eng.: Lesser Mottled Grasshopper.

This species is divided into two subspecies. The nominate ssp. *stigmaticus stigmaticus* Ramb. is restricted to the Iberian Peninsula and Morocco whereas the subspecies found in the rest of Europe is named *stigmaticus faberi* Harz, 1975.
The coloration may vary between green and brown. Fore wings narrow, extending approximately to the tip of the abdomen or a little further in the male and a little less in the female. Cubitus-1 and cubitus-2 of fore wings separate. A small species. Length: ♂ 11-15 mm; ♀ 15-20 mm.

Distribution. *S. s. faberi* does not occur in Denmark or Fennoscandia. The nearest occurrence is in Holstein, where it has been recorded in several localities near Mölln, south of Lübeck. — In the British Isles, occurs only on the Isle of Man. France, Central Europe, Hungary, Yugoslavia, Rumania and the European part of USSR.

Biology. Inhabits drier and warmer places than the previous species. It can be found in e.g. moorland and clearings in (planted) woodland.

Genus *Omocestus* Bolivar, 1878

Omocestus Bolivar, 1878: 427, Ann. Soc. Esp. 7: 427.
   Type-species: *Gryllus viridulus* Linnaeus, 1758.

Foveolae narrow and rectangular. Antennae filiform. Leading edge of fore wings straight, with no widening towards the base. Medial area of fore wings not extended. Side keels on pronotum more or less angularly introflexed (greatest width between them two or three times as wide as at the narrowest spot). Valves of ovipositor non-denticulate.

The fastigium verticis may have a median keel (Fig. 56A) in some species, though it may be difficult to see if the anterior portion of the head is worn. The point can be decided by looking to see whether the margins of the foveolae are worn down. A median keel can always be felt with a mounting needle and/or seen in a bright light shone from the side, or under powerful magnification.

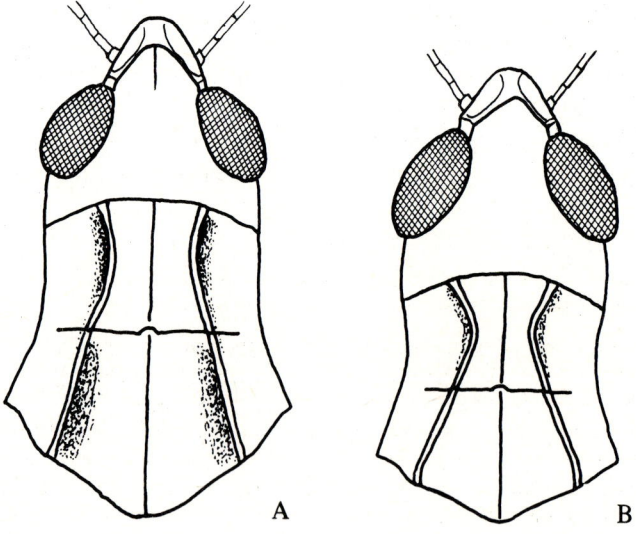

Fig. 56. Head and pronotum of A: *Omocestus viridulus* (L.) and B: *O. haemorrhoidalis* (Charp.).

Key to species of *Omocestus*

1 Side keels on pronotum curved roundly inwards in the prozona (Fig. 56A) 2
– Side keels on pronotum curved angularly inwards in the
  prozona (Fig. 56B) .... 34. *haemorrhoidalis* (Charpentier)
2 (1) Median furrow present on vertex (Fig. 56A). Valves of ovipositor same length as subgenital plate (Fig. 57A). Palps
  uniform in colour .... 32. *viridulus* (Linnaeus)
– No median furrow on vertex. Valves of ovipositor shorter
  than subgenital plate (Fig. 57B). Distal segment of palps
  more or less white in colour .... 33. *ventralis* (Zetterstedt)

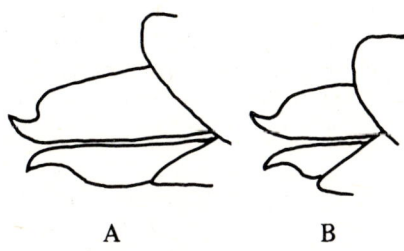

Fig. 57. Ovipositor of A: *Omocestus viridulus*
(L.) and B: *O. haemorrhoidalis* (Charp.).

A   B

32. **Omocestus viridulus** (Linnaeus, 1758)
    Figs. 56A, 57A, 58-60.

Gryllus viridulus Linnaeus, 1758, Syst. Nat. ed. 10: 433.
Eng.: Common Green Grasshopper; Dan.: Lynggræshoppen; Fin.: Niittyheinäsirkka;
Norw.: Grønn markgrashoppe; Sw.: Gröna ängsgräshoppan.

Most commonly olive-green in colour, with brown to yellow sides. Completely green,

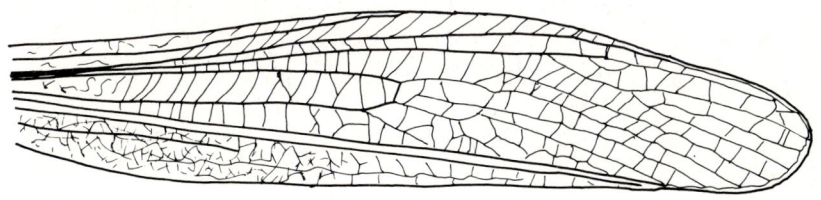

Fig. 58. Fore wing of *Omocestus viridulus* (L.) ♂.

brown or yellow individuals are, however, also found. Eyes small. Side keels on pronotum curved roundly inward in the prozona. Vertex with longitudinal furrow. Valves of
ovipositor same length as subgenital plate. Fore wings extend to or beyond the ends of
the femora. Hind wings smoky in colour. Numbers of stridulatory pegs: ♂ 100-140; ♀
90-135. Length: ♂ 13-15 mm; ♀ 20-24 mm.

Distribution. Generally distributed throughout Denmark and Fennoscandia,
though absent from the northern parts of Sweden, Norway and Finland. — From
Western Europe (the British Isles to N.Spain) through Central Europe to N.Italy, the
Balkans, USSR (to Kazakhstan and Siberia) and Mongolia.

Biology. Mainly inhabits damp places, but is also found in drier localities. Frequently found with *Metrioptera brachyptera.* May be found in clearings in woods, in
meadows, beside ditches and on damp heathland. The male flies vigorously on hot
days. The egg capsules ('pods') are 7-12.5 mm in length, their surface consisting of a
hardened secretion. A pod may contain up to ten eggs, which are diagonally placed to
the longitudinal axis of the capsule. The eggs are 4-4.6 mm in length, a dirty whitish
brown in colour and with a fine structure. The capsules are deposited at the base of
clumps of grass. This species develops very early in Denmark and southern Fennoscandia, the first nymphs emerging in May if it is a warm Spring.

Stridulation. The sequence may last for a very short time or up to a whole minute.
The first chirps in a sequence are weak, getting up to full volume after a few seconds
and sounding like ticking. A sequence terminates very abruptly. The number of chirps
is between 9 and 10 per second. Each chirp consists of a long faint upstroke syllable
(a in fig. 60), produced by the upward movement of the hindlegs, and a shorter but
louder syllable (b in fig. 60), produced during their downward movement (Elsner,
1974).

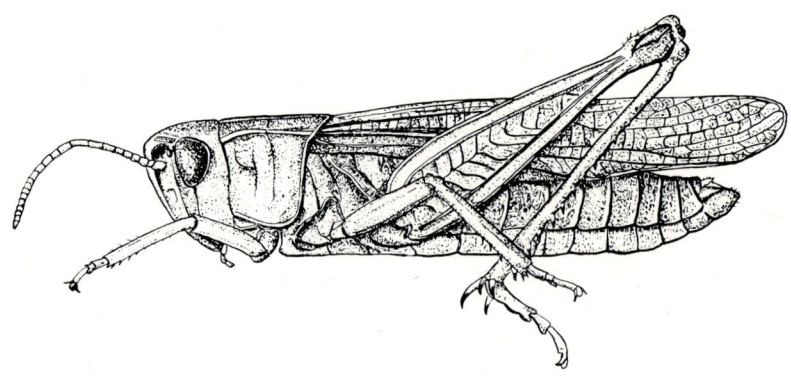

Fig. 59. *Omocestus viridulus* (L.) ♂.
Length 13-15 mm. (From Harz, 1975).

a　b

Fig. 60. Oscillogram of the song of *Omocestus viridulus* (L.). — DK, NEZ: Grib Skov, August 1967. Scale: 0.25 sec.

33. **Omocestus ventralis** (Zetterstedt, 1821)
　　Fig. 61.

Gryllus ventralis Zetterstedt, 1821, Orth. Svec.: 89.
Gryllus rufipes Zetterstedt, 1821.
Eng.: Woodland Grasshopper; Sw.: Rödgumpgräshoppan.

Coloration approximately the same as in the previous species, though the darker forms are commoner. Eyes large. Side keels on pronotum curved roundly inwards in the prozona. Vertex lacks a longitudinal furrow. Valves of ovipositor shorter than subgenital plate. Terminal segment of palps completely white or white with dark rings. Rest of palp dark. Fore wings extend beyond or to end of hind femora. Hind wings smoky in colour. Numbers of stridulatory pegs: ♂ 90-130; ♀ 90-120. Length: ♂ 12-17 mm; ♀ 18-20 mm.

Distribution. Not found in Denmark. — Found at several sites in southern Sweden, extending as far north as Uppland, Västmanland and Värmland. — In Norway, occurs sporadically in southerly coastal areas. — Finland: this species is only known from the Satakunta Province (St) on the Gulf of Bothnia. — South England, Wales, France, Central Spain, Portugal, Algeria, Central and East Europe, the Balkans, Turkey and USSR to South Siberia, the Northern Caucasus and Kazakhstan.

Biology. Prefers dry, warm localities covered with e.g. grass, or found in association with *Calluna* and *Erica* near sunny cliffs.

Stridulation. Very much like *O. viridulus,* though the sequence is shorter. Near Kristiansand, in S.Norway, a sequence lasting nine seconds has been recorded, and one lasting six seconds in Scania, Sweden (Wallin, 1979). The first chirps are weak, but rapidly attain full volume. The number of chirps is from about 20 per second (Kristiansand) to about 10 per second (Scania). Each chirp consists of a long upstroke syllable, most probably produced during the upward movement of the hindlegs, and a short loud downstroke syllable, most probably produced during their downward movement.

93

Fig. 61. Oscillogram of the song of *Omocestus ventralis* (Zett.). — N, AAy: near Kristiansand, August 1978. Scale: 0.25 sec.

34. *Omocestus haemorrhoidalis* (Charpentier, 1825)
Figs 56B, 57B, 62.

Gryllus haemorrhoidalis Charpentier, 1825, Hor. Ent.: 165.
Sw.: Alvargräshoppan.

Commonly olive-green dorsally, with yellow to brown sides. Completely green or yellowish brown forms also occur, however. In Sweden this species is generally dark grey. In the anterior region of the fore wings the venation is brown with individual brown spots whilst the venation in the posterior region is green (in green individuals) or brown (in brown ones). Fore wings do not extend beyond ends of hind femora. Hind wings colourless. Eyes large. Side keels on pronotum angularly introflexed in prozona. No longitudinal furrow in vertex. Valves of ovipositor shorter than subgenital plate. Distal portion of abdomen pink to reddish yellow superiorly in the male. Numbers of stridulatory pegs: ♂ 120-165; ♀ 100-150. Length: ♂ 11-13 mm; ♀ 16-19 mm.

Distribution. Occurs sporadically in Denmark: Mols, near Århus (EJ); the island of Læsø (NEJ); and in S.Sweden: central Skåne; on Kinnekulle in Västergötland; on Öland and Gotland. Found in several localities on the Mols peninsula and the islands of Öland and Gotland. — In addition this species has been found at several sites in the area south and west of Lübeck. This discontinuous distribution suggests the presence of a relict from the Boreal period (Ander, 1949). The total distribution includes France, W.Spain, Central Europe, Italy, Hungary, Rumania, Bulgaria, USSR, mountains of Central Asia, and Korea.

Biology. Moorlands and dry graminetae. In Denmark and N.Germany it is to be found on sunny, sandy sites, and in Sweden, on heat-absorbent chalk cliffs. It is frequently found with *Chorthippus brunneus, Ch. biguttulus* and *Myrmeleotettix maculatus.*

Stridulation. A sequence has been recorded on the Mols peninsula in Jutland lasting 1.5 to about 3 seconds, and in Västergötland up to about three seconds (Wallin, 1979).

The chirps start weakly, rapidly attaining their full metallic sound, being produced at a rhythm of 10-20 chirps per second. Each chirp consists of several syllables.

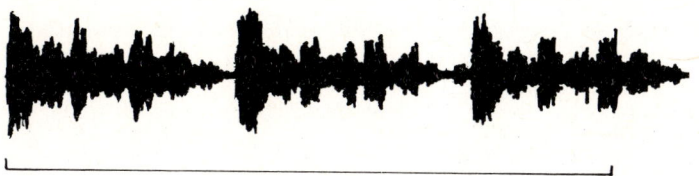

Fig. 62. Oscillogram of the song of *Omocestus haemorrhoidalis* (Charp.). — DK, EJ: Strandkær, Mols, August 1978. Scale: 0.25 sec.

Genus *Stauroderus* Bolivar, 1898

Stauroderus Bolivar, 1898, Ann. Sci. Nat. Porto 4: 424.
   Type-species: *Oedipoda scalaris* Fischer-Waldheim, 1846.

Foveolae narrow and rectangular. Antennae filiform. Leading edge of fore wings slightly enlarged at the base, cubitus-1 and cubitus-2 being fused. The medial and costal areas of the fore wings are very highly developed in the male, with regular parallel veins. These broad areas make the entire fore wing broad. Side keels on pronotum slightly angularly introflexed. Valves of ovipositor non-denticulate.

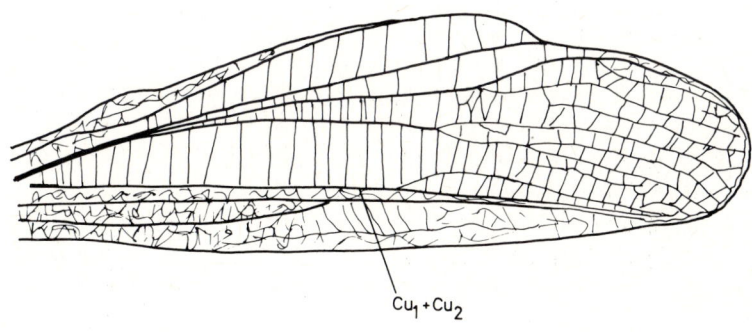

Fig. 63. Fore wing of *Stauroderus scalaris* (F.-W.) ♂.

95

35. *Stauroderus scalaris* (Fischer-Waldheim, 1846)
Fig. 63.

Oedipoda scalaris Fischer-Waldheim, 1846, Orth. Ross.: 317.
Dan.: Den skærende græshoppe; Sw.: Den skärrande gräshoppan.

Coloration may vary from brown to reddish yellow. Hind tibia red to reddish yellow; fore wings brown, with brightly coloured brown tips. Length: ♂ 18-22 mm; ♀ 23-29 mm (measurements of individuals from The Alps).

Distribution. Sweden: only found on the north of the island of Öland. Must be regarded as a relict from the Continental Boreal period, since its main area of distribution is the Russian steppes. — Found at various localities in the mountains of Central and Southern Europe.

Biology. On dry localities with sparse, semi-withered grass in Summer.

Stridulation. The sequence lasts about 30 seconds. The individual chirps (about 2 per sec.) are very sharp, terminating with a short click. The sound is very characteristic, resembling sawing (Wallin, 1979).

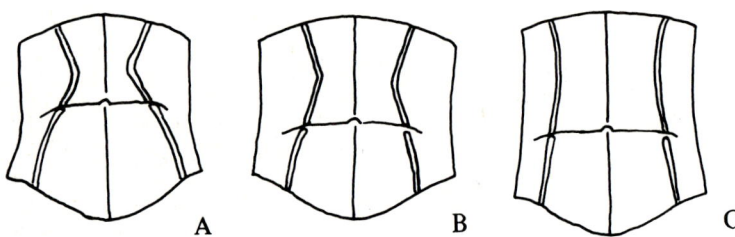

Fig. 64. Pronotum of A: *Chorthippus brunneus* (Thnbg.); B: *Ch. apricarius* (L.) and C: *Ch. parallelus* (Zett.).

Genus *Chorthippus* Fieber, 1852

Chorthippus Fieber, 1852, *in* Kelch: Orth. Oberschlesiens: 1.
Type-species: *Acridium albomarginatus* De Geer, 1773.

Foveolae narrow and rectangular. Antennae filiform. Leading edge of fore wings slightly widened at base (though it may be weakly developed or even absent in *Ch. apricarius* ♀ and *Ch. albomarginatus* ♂). Cubitus-1 and cubitus-2 separate, except in *Ch. apricarius,* where they are separate only until a short distance beyond their origin. Medial area not powerfully extended (extended to a certain degree in *Ch. apricarius*). Costal area prominent in the male, crowding out the precostal area, which thus becomes shortened. In the female, the costal area of the fore wings is not so extensive, the precostal area being long and extended towards the wingtips. Valves of ovipositor non-denticulate.

Some of these species may be difficult to separate, particularly *Ch. brunneus, Ch. biguttulus* and *Ch. mollis,* in which the males may be difficult to identify and the females virtually impossible unless the number of stridulatory pegs is known.

This genus is plentifully represented in Europe.

Key to subgenera and species of *Chorthippus*

1   Side keels on pronotum angularly introflexed in the prozona and highly divergent in the metazona (Figs 64A & B) . . . . Subgenus *Glyptobothrus* 2
  –   Side keels on pronotum straight or curved slightly inwards (Fig. 64C) . . . . Subgenus *Chorthippus* s. str. 7

2 (1) Prozona shorter than metazona (Fig. 64A). Tympanum two-thirds covered (Fig. 35C) 3
  –   Prozona longer than or nearly same length as metazona (Fig. 64B). Tympanum one third covered (Fig. 35B) 5

3 (2) Fore wings wide in male, the costal area being considerably widened and the subcostal area slightly widened (Fig. 72). Numbers of stridulatory pegs 80-130 in the male and 75-120 in the female . . . . 38. *biguttulus* (Linnaeus)
  –   Fore wings of normal width. 4

4 (3) Fore wings in male three quarters as wide at the tip as at the widest place (Fig. 65). Length of fore wings 13-15 mm in male and 19-21 mm in female. Numbers of stridulatory pegs 45-90 in male and 50-90 in female . . . . 36. *brunneus* (Thunberg)
  –   Fore wings only half as wide at the tip as at the widest place (Fig. 69). Length of fore wings 10-13 mm in male and 13-15 mm in female. Numbers of stridulatory pegs 100-135 in male and 85-120 in female . . . . 37. *mollis* (Charpentier)

5 (2) Cubitus-1 and cubitus-2 of fore wing fused, except for initial sector closest to origin. Medial area widened (Fig. 75). . . . . 39. *apricarius* (Linnaeus)
  –   Cubitus-1 and cubitus-2 of fore wing separate 6

6 (5) Fore wings in male extend beyond tip of abdomen and in female only to tip of abdomen. Subcosta straight . . . 40. *vagans* (Eversmann)
  –   Fore wings in male extend to tip of abdomen; and are shorter in the female. Subcosta curved. . . . . 41. *pullus* (Philippi)

7 (1) Both pairs of wings well-developed 8
  –   Hind wings shorter than fore wings in both sexes. Fore wings extend to tip of abdomen or less in male, and are

flap-like in female 9

8 (7) Side keels practically straight and only slightly curved outwards posteriorly. Sigmoid upward curvature in radius, and
media down-curved, resulting in radial area widening out
near wingtip (Fig. 80) . . . . 42. *albomarginatus* (De Geer)

–   Side keels only slightly introflexed. Radius not sigmoid up-
wardly curved (Fig. 82) . . . . 43. *dorsatus* (Zetterstedt)

9 (7) Transverse sulci on pronotum situated a few mm posterior
to the mid-point (fig. 64C). Hind wings in male reach to approximately the middle of the fore wings. Valves of ovi-
positor short . . . . 44. *parallelus* (Zetterstedt)

–   Transverse sulci in middle of pronotum. Hind wings in
male extend approximately to the outer quarter of the fore
wings. Valves of ovipositor long . . . . 45. *montanus* (Charpentier)

Subgenus *Glyptobothrus* Chopard, 1951

Glyptobothrus Chopard, 1951, 192.
   Type-species: *Gryllus binotatus* Charpentier, 1825.

Side keels of pronotum angularly incurved (introflexed).

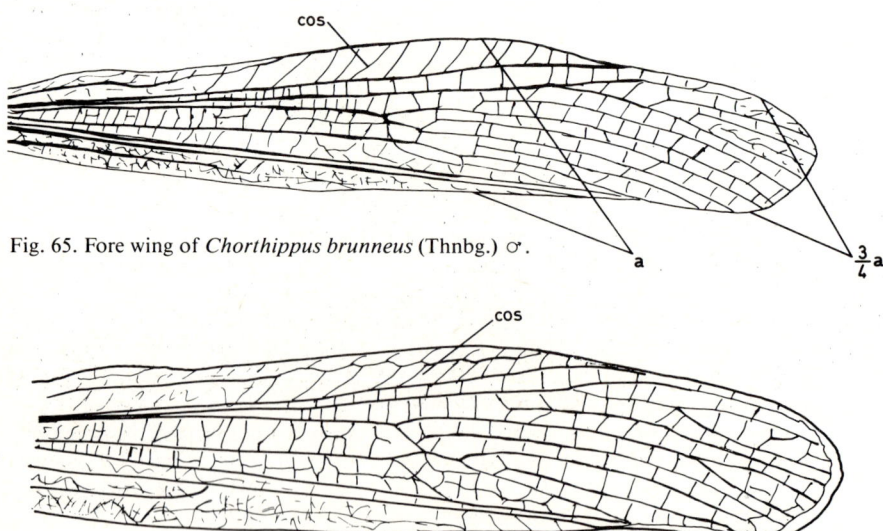

Fig. 65. Fore wing of *Chorthippus brunneus* (Thnbg.) ♂.

Fig. 66. Fore wing of *Chorthippus brunneus* (Thnbg.) ♀.

36. *Chorthippus (Glyptobothrus) b. brunneus* (Thunberg, 1815)
Figs 35C, 49, 64A, 65-68.

Gryllus brunneus Thunberg, 1815, Mém. Acad. Sci. St. Pétersbg. 5: 249.
Gryllus bicolor Charpentier, 1825, Hor. Ent.: 171.
Chorthippus bicolor brevis Klingstedt, 1939, J. Gen. 37: 393.
Eng.: Common Field Grasshopper; Dan.: Den almindelige markgræshoppe; Fin.:
Ketoheinäsirkka; Norw.: Gråbrun markgrashoppe; Sw.: Den vanliga fältgräshop-
pan.

Wide variations in coloration possible: green, yellow, brown, red, grey and black. Individuals may be uniform in colour or more or less mottled. Side keels angularly introflexed. Transverse sulcus on pronotum just anterior to the mid-point, making the
prozona shorter than the metazona. Fore wings narrow and rather variable in length,
though they usually extend beyond the ends of the hind femora. The costal area (cos)
in the male is more or less prominent and may in some specimens resemble that in *C.*
biguttulus. Costa usually straight where it meets the subcosta or the precostal area,
though there may sometimes be a small bend in it. Fore wings in male three quarters
as wide at the tip as at the widest spot. Tympanum two-thirds obscured (Fig. 35C).
Tip of abdomen frequently red in males. Numbers of stridulatory pegs: ♂ 45-90; ♀
50-70. Length: ♂ 13-17 mm; ♀ 19-24 mm. Length of fore wings: ♂ 13-15 mm; ♀
19-21 mm.

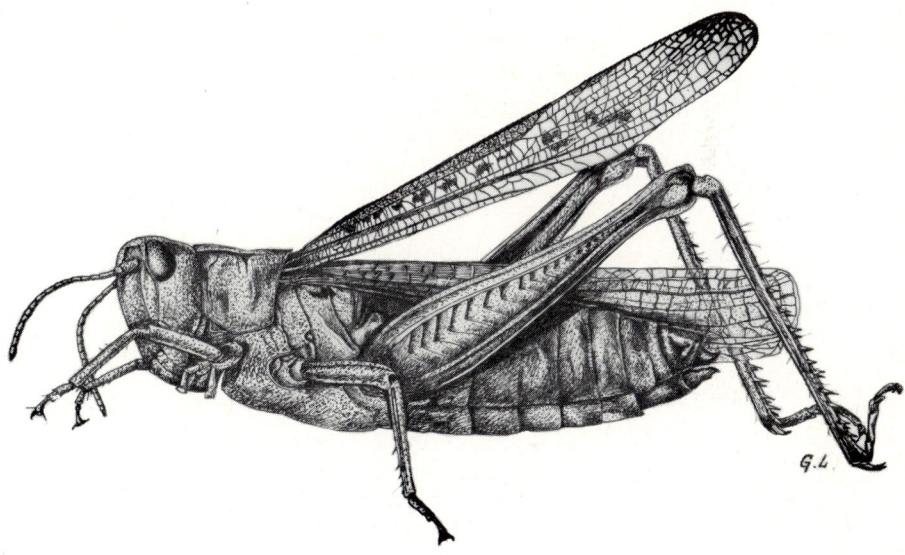

Fig. 67. *Chorthippus brunneus* (Thnbg.) ♀. Length 19-24 mm.

Distribution. Widely and generally distributed throughout Denmark and the greater part of Sweden, Norway and Finland, although absent from the northernmost parts of these countries as well as from the Scandinavian mountain chain. — Most parts of W., C., S. & E. Europe; Asia Minor; N. Africa; eastwards over European USSR, Kazakhstan, the Caucasus, Irak and Iran to Mongolia and N.China.

Biology. This species is beyond any doubt the most tolerant of all grasshoppers. It can be found in a wide variety of biotopes, except for the very driest and the very wettest. It lives, for example, in and beside ditches, on the "white" and the "grey" dunes, on slopes and hillsides, on the edges of woods and on moorlands — but not in the mountains. Very active on hot days. It can fly several metres at a time, and alter its direction while in flight, the male being a better flier than the female.

The egg capsules are laid in the soil, just below the surface. The eggs hatch May-June (more commonly June). Imagines are seen in July-August and may survive on into October. The pods are 11.5-16 mm in length, cylindrical or elongated with straight or curved sides and with a round lid at the top. The surface of the pod consists of hardened secretion and soil. A pod may contain up to 14 eggs, which are diagonally placed to the longitudinal axis of the pod. The eggs are 3.9-4.5 mm in length, with a smooth surface.

Stridulation. Stridulation in this species comprises a series of short chirps, each of which lasts less than half a second and which may be repeated at various rhythms but which, at optimum temperatures, are repeated at one to two second intervals. An entire sequence usually consists of from six to twelve chirps. Each chirp consists of several syllables.

Note. The "subspecies" *Chorthippus bicolor brevis* Klingstedt, 1939, is not accepted here. It was described as having a broader and shorter fore wing, with a radius being closer to the wing base ($3/5$ or less, as against $2/3$ or more in *C. brunneus*). The number of stridulatory pegs varies between 102 and 143 (n = 10). Stridulation in *brevis* does not differ from that it the nominate subspecies.

Fig. 68. Oscillogram of the song (one chirp) of *Chorthippus brunneus* (Thnbg.). — DK, NEZ: Tisvilde, August 1979. Scale: 0.25 sec.

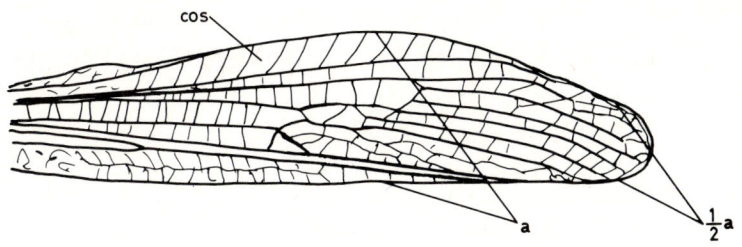

cos

Fig. 69. Fore wing of *Chorthippus mollis* (Charp.) ♂.

37. *Chorthippus (Glyptobothrus) m. mollis* (Charpentier, 1825)
Figs 69-71.

Gryllus mollis Charpentier, 1825, Hor. Ent.: 164.

Colours may be variable, much as in *C. brunneus*. Side keels angularly introflexed. Transverse sulcus on pronotum anterior to the mid-point, the prozona thus being shorter than the metazona. Fore wings usually extend a little beyond ends of hind femora. Fore wings narrow; wing tips in male half the width of the widest spot. Tympanum two-thirds obscured. Tip of abdomen frequently red in colour (Nordic specimens). Numbers of stridulatory pegs: ♂ 100-135; ♀ 85-125. Length: ♂ 12.5-14 mm; ♀ 17-18 mm. Length of fore wings: ♂ 10-13 mm; ♀ 13-15 mm.

Distribution. In this area, only found in Denmark: some in a few isolated localities in SJ and some on coastal slopes in the northern part of the island of Samsø (EJ). Also in isolated localities in Schleswig-Holstein. — Also occurs south of L. Ladoga in the USSR, but is not found in Sweden, Norway or Finland. — The total distribution covers C. and E. Europe, Spain, Italy and the Balkans; eastwards over the European USSR to the Caucasus, Kazakhstan, C.Asia, Siberia, Turkey and N.Iran.

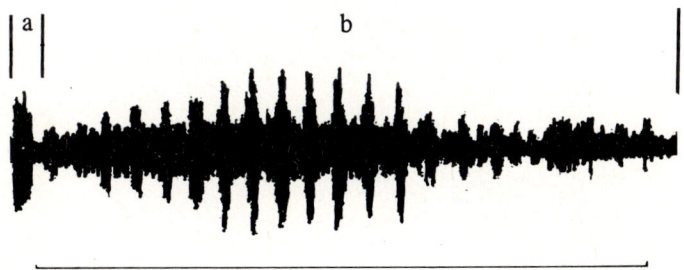

a   b

Fig. 70. Oscillogram of the song of *Chorthippus mollis* (Charp.). — D, Holstein: Göttin, August 1967. Scale: 0.25 sec.

Fig. 71. Diagram of the song of *Chorthippus mollis* (Charp.).

Biology. Inhabits rather drier areas than *brunneus,* e.g. inland dune areas and sandy slopes.

Stridulation. The sequence lasts from 15 to 30 seconds and consists of a series of short chirps which virtually merge into each other and are produced at the rate of 3-5 chirps per second. The initial chirps are weak, but they rapidly attain full strength, beginning as a quivering sound and terminally developing into a loud snarl. The sequence terminates with 2-4 weak chirps. The chirps produced by one of the hindlegs begin with a loud downstroke syllable (a) which can be identified careful listening. Thereafter follows a vibratory phase produced by a series of upward and downward movements (b) (Elsner, 1974).

38. ***Chorthippus (Glyptobothrus) b. biguttulus*** (Linnaeus, 1758)
   Figs 72-74.

Gryllus biguttulus Linnaeus, 1758, Syst. Nat. ed. 10: 433.
Fin.: Ahoheinäsirkka; Sw.: Slåttargräshoppan.

Colours vary much as in *Ch. brunneus.* Side keels angularly introflexed. Transverse sulcus on pronotum slightly anterior to mid-point, the prozona thus being shorter than the metazona. Fore wings usually extend a little way beyond the end of the hind femora. Costal and subcostal areas of fore wings (cos and sub) strongly developed in the male, resulting in a broad wing. Costa usually bends at point at which it meets subcosta after the precostal area. Fore wings in male half as wide at the tip as at their widest extent. Tympanum two-thirds obscured. Tip of abdomen frequently red in male.

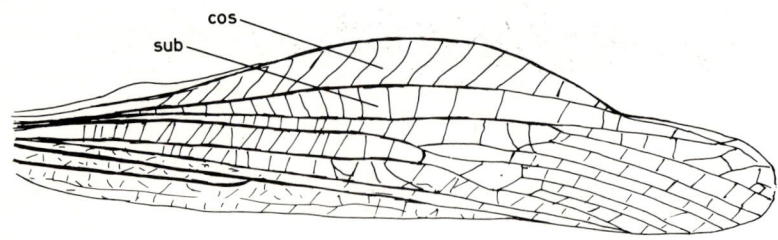

Fig. 72. Fore wing of *Chorthippus biguttulus* (L.) ♂.

Fig. 73. Oscillogram of the song of *Chorthippus biguttulus* (L.). — DK, EJ: Strandkær, Mols, August 1978. Scale: 0.25 sec.

Numbers of stridulatory pegs: ♂ 80-130; ♀ 75-120. Length: ♂ 13-15 mm; ♀ 17-22 mm.

Distribution. Denmark: scattered occurrence, except in the northwestern and northern parts of Jutland. — Sweden: from Sk. to Dlr. and Gstr., but also further north in the lowlands of Ång., Vg. and Nb. Absent from Gtl. — Norway: only in districts round Oslo Fjord; south to Arendal. — Finland: from Al to ObS, though no records from Oa and Om. — Is distributed over most of Europe except for the British Isles. In the south to NE.Spain and the Alps; in the east to Hungary, Rumania and Bulgaria.

Biology. Inhabits light, dry, frequently sandy and sunny spots where such plants as *Trifolium arvense, Galium verum* and *Agrostis* spp. grow. It can for example be found on dry-stane dykes, the steep sides of cuttings and in clearings in woods. Its biology is reminiscent of that of *Ch. brunneus*.

Stridulation. The second order sequences last 8-9 seconds and consist of three first order sequences. The first sequence lasts about two and a half seconds; after about one second's interval the next sequence, lasting about one and a half seconds, follows and finally, after yet another interval lasting about two and a half seconds, the last sequence, of about the same duration as the previous one: some one and a half seconds. Each first order sequence commences with a certain number of weak chirps produced at a rhythm of 20-30 chirps per second, with a loud, metallic noise. Each chirp cousists of several syllables.

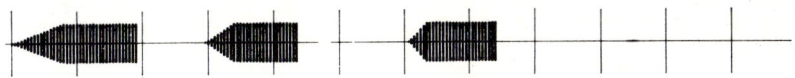

Fig. 74. Diagram of the song of *Chorthippus biguttulus* (L.) ♂.

103

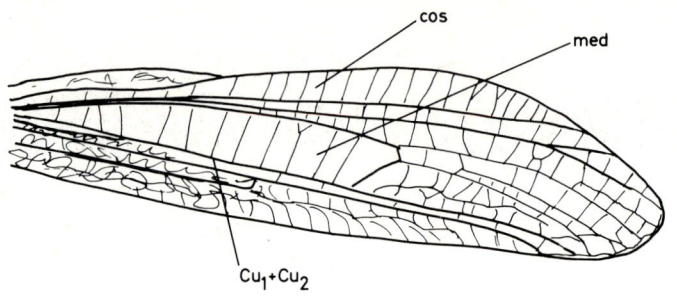

Fig. 75. Fore wing of *Chorthippus apricarius* (L.) ♂.

39. *Chorthippus (Glyptobothrus) apricarius* (Linnaeus, 1758)
Figs 35B, 64B, 75-77.

Gryllus apricarius Linnaeus, 1758, Syst. Nat. ed. 10: 433.
Sw.: Solgräshoppan.

Usually yellow to brown in colour, with more or less extensive darker portions. Side ke-els angularly introflexed. Transverse sulcus on pronotum sited a little way posterior to the mid-point, the prozona thus being longer than the metazona. Fore wings in female rarely extend beyond the tip of the abdomen whilst in the male they reach the end of the hind femora. The restricted widening at the base of the fore wings is not as highly developed as in the other *Chorthippus* species in which, particularly in the female, it is weakly developed or even absent. Cubitus-1 and cubitus-2 of the fore wings fused ex-cept for the initial section after the origin. Medial and costal areas of fore wings wide-ned, broadening the wings. These areas are especially well developed in the male, bea-ring regular, parallel veins, whilst in the female they are not so well developed and have irregular venation. Tympanum one third obscured. Numbers of stridulatory pegs: ♂ 130-200; ♀ 135-175. Length: ♂ 14-16 mm; ♀ 18-22 mm.

Distribution. In this area, in Denmark and Sweden only, in a belt stretching from

Fig. 76. Oscillogram of the song (almost one chirp) of *Chorthippus apricarius* (L.). — DK, LFM: Berritskov, September 1969. Scale: 0.25 mm.

Fig. 77. Diagram of the song of *Chorthippus apricarius* (L.).

Schleswig-Holstein, southern and eastern Jutland, Funen, Lolland, Falster, South Zealand, Skåne, Halland, Blekinge and Småland to the islands of Öland og Gotland. — France, Central Europe, Hungary, Rumania, Bulgaria, USSR, Mongolia and N. China.

Biology. May be very common by the roadside and in places where grass grows permanently with a slight tendency towards drought.

Stridulation. The sequence may vary, lasting right up to 45 seconds. Commences with a certain number of very weak chirps. The fact that this species has started stridulating is best determined by observing the hind legs moving up and down. After about five seconds the "song" has reached full volume, sounding like "seeD-seeD-seeD?" with a three to five chirps second rhythm. Their "song" is highly characteristic and cannot be mistaken for that of any other species.

40. *Chorthippus (Glyptobothrus) vagans* (Eversmann, 1848)
   Figs 78, 79.

Oedipoda vagans Eversmann, 1848, Addit. Orth. Ross.: 12.
Eng.: Heath Grasshopper.

Colours may vary in the yellow, brown or grey range. The side keels are angularly introflexed. Transverse sulcus on pronotum situated just posterior to the mid-point, the prozona thus being longer than the metazona. Fore wings in male extend a little beyond the tip of the abdomen but do not reach the end of the hind femora; a bit shorter in the female. Costal area of fore wing in male only slightly protuberant; cubitus-1 and

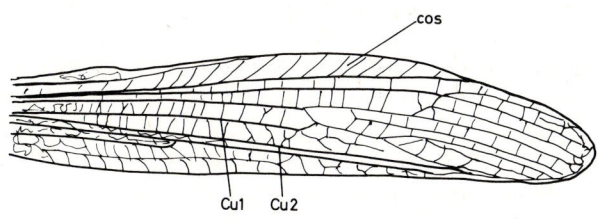

Fig. 78. Fore wing of *Chorthippus vagans* (Ev.) ♂.

105

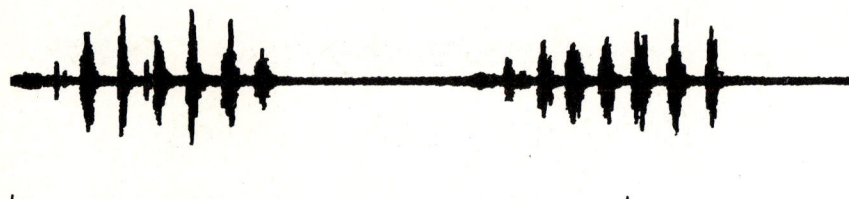

Fig. 79. Oscillogram of the song (2 chirps) of *Chorthippus vagans* (Ev.). — DK, NEJ: Skagen, August 1978. Scale: 0.25 sec.

cubitus-2 widely separated. Tympanum one third obscured. Numbers of stridulatory pegs: ♂ 120-180; ♀ 120-160. Length: ♂ 12-16 mm; ♀ 16-22 mm.

Distribution. In this area, only found in Denmark, where it is common round The Skaw (NEJ) and in forestry plantations on dunes, on moorlands and in the dunes, e.g. at Råbjerg Mile. Also found on a dune near Lübeck in Holstein. — A very restricted occurrence in S.England as well. Recorded from various European countries, south to the Mediterranean and east into USSR. Many records require verification.

Biology. On the sandy soil of the grey and white dunes, on heathland and in clearings in forestry plantations. It may be found on bare sandy soil but it is commoner in heather and drought prone grasses. It is frequently found in association with *Myrmeleotettix maculatus*.

Stridulation. The sequence lasts about three seconds (The Skaw, Jutland) and consists of a series of short chirps (six chirps per second). Each chirp consists of several syllables.

41. *Chorthippus (Glyptobothrus) pullus* (Philippi, 1830)

Gryllus pullus Philippi, 1830, Orth. Berol.: 38.

Yellowish to red brown in colour. Side keels angularly introflexed. Prozona longer than or almost as long as metazona. Fore wings in male extend to tip of abdomen. Costal area very much extended and medial area somewhat extended. Subcosta curved. Fore wings in female extend to middle of abdomen. Length: ♂ 12-14 mm; ♀ 19-20 mm (Central European specimens).

Distribution. In this area, only found in the Leningrad region and at L.Ladoga in the USSR. — Occurs sporadically in France, Central Europe, Hungary, Rumania; in the European USSR south to the Caucasus.

Biology. On warm, dry spots, both in lowland country and in mountains. Very local.

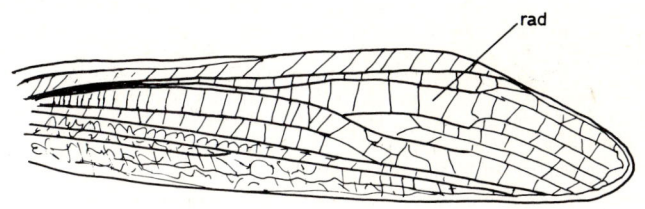

Fig. 80. Fore wing of *Chorthippus albomarginatus* (De Geer).

Subgenus *Chorthippus* s. str.

Side keels of pronotum straight or curved slightly inwards.

42. *Chorthippus (Ch.) a. albomarginatus* (De Geer, 1773)
Figs 80, 81.

Acrydium albomarginatus De Geer, 1773, Mém. Ins. 3: 480.
Eng.: Lesser Marsh Grasshopper; Dan.: Strandengsgræshoppen; Fin.: Rantaheinä-sirkka; Sw.: Strandängsgräshoppen.

Uniform or more or less mottled yellow, brown or green coloration. The side keels are practically straight and only very slightly outwardly curved posteriorly. Length of fore wings may vary, extending beyond the tip of the abdomen or not reaching so far. Widening at base of fore wings not as pronounced as in other *Chorthippus* species; may even be absent. Radius and media of fore wings run close together from their origin to the middle of the wing, from which point onwards the radius makes a sigmoid curve forwards and the media bends backwards, resulting in the medial area becoming wide, with regular parallel venation. This widening of the radial region is most marked in the male. The costal area of the fore wings is not particularly prominent in the male. Tympanum one-third obscured. Numbers of stridulatory pegs: ♂ ♀ 100-130. Length: ♂ 12-16 mm; ♀ 17-20 mm.

Distribution. Very common near coasts; rarer inland. Found throughout Denmark and Schleswig-Holstein. — In Sweden, it has been found as far north as Central Sweden, and in Norway it is found round the south coast and as far up as Bergen. — In Finland it is common in the south, including Alandia (The Åland Islands) and the islands of the Gulf of Finland. It is found on the Finnish side of the Gulf of Bothnia and one must presume that it is present on the Swedish side as well. Round L. Ladoga in the Soviet Union. — Ireland, Wales, England, Central Europe, France, Spain, Italy, Rumania, Turkey, USSR and Central Asia.

Biology. Inhabits the grey dunes, the white dunes and permanent stands of grass, especially near coasts. Has also been found in littoral meadows and in bogs, frequently in association with *Conocephalus dorsalis*.

The egg capsules are 7.5-12 mm in length, irregular in shape, and with no lid. The surface of the capsule consists of agglutinated secretion. It may contain up to ten eggs which are diagonally situated to the longitudinal axis of the capsule. The eggs are 3.8-4.5 mm long, with a smooth surface. The pods are deposited at the base of tufts of grass.

Stridulation. Stridulation in this species consists of a series of short, whirring chirps lasting a little under a second and repeating at approximately two second intervals. The entire sequence consists of some two to six chirps. Each chirp consists of several syllables.

Fig. 81. Oscillogram of the song of *Chorthippus albomarginatus* (De Geer). — DK, NEZ: Lynæs, August 1967. Scale: 0.25 sec.

43. *Chorthippus (Ch.) dorsatus* (Zetterstedt, 1821)
Figs 82, 83.

Gryllus dorsatus Zetterstedt, 1821, Orth. Svec.: 82.
Sw.: Syd-ängsgräshoppan.

Coloration may vary. The most usual colours are yellow, brown or green. Side keel slightly introflexed. Fore wings extend nearly to ends of hind femora; radial area normal in size. Tympanum one-third obscured. Numbers of stridulatory pegs: ♂ 90-150; ♀ 100-140. Length: ♂ 14-18 mm; ♀ 19-26 mm.

Distribution. Scattered in Denmark (not in NWJ & NEJ) and South Sweden: Sk.,

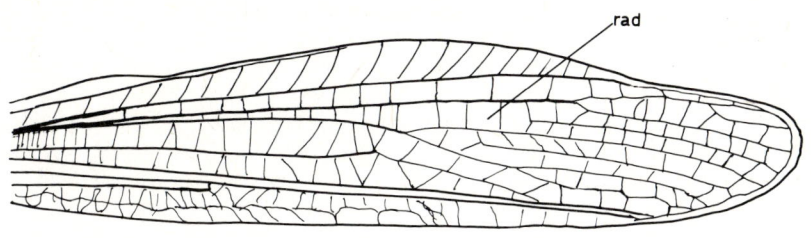

Fig. 82. Fore wing of *Chorthippus dorsatus* (Zett.).

Bl., Sm., Öl., Ög., Boh. and Nrk. — Not found in Norway or Finland. — Common in S. Holstein. — France, C. Europe, N.Spain, Italy, Yugoslavia, Rumania, Bulgaria and USSR.

Biology. On drier sites. It has been found on the lighter soils with sparse vegetation, on grassy slopes and in heather.

Fig. 83. Oscillogram of the song (one sequence with several chirps) of *Chorthippus dorsatus* (Zett.). — DK, NEZ: Tisvilde, August 1979. Scale: 0.5 sec.

Stridulation. The sequence in Swedish individuals caught in Öland (Wallin, 1979) last about one and a half seconds. In specimens caught in Tisvilde, Denmark, duration is about two seconds. The sequences are repeated at a few seconds' intervals. The first four or five chirps are short, the initial ones commencing weakly. The terminal part of the sequence turns into a long, sharp, tinkling or jingling sequence lasting seven tenths of a second (Öland) or a little less than one second (Tisvilde). In the first part of the sequence the individual chirps have a tendency to merge, though one can hear them apart, whilst those in the final long sequence merge into one another.

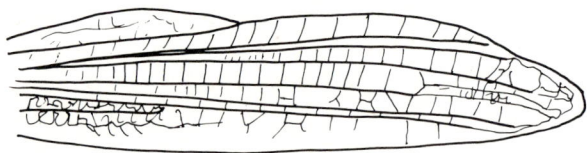

Fig. 84. Fore wing of *Chorthippus parallelus* (Zett.).

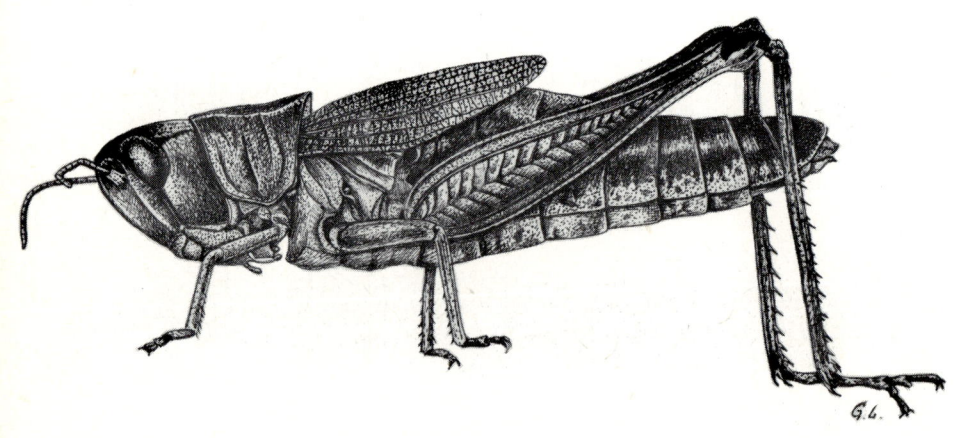

Fig. 85. *Chorthippus parallelus* (Zett.) ♀.
Length: 16-23 mm.

44. *Chorthippus (Ch.) p. parallelus* (Zetterstedt, 1821)
Figs 64C, 84-86.

Gryllus parallelus Zetterstedt, 1821, Orth. Suec.: 85.
Acrydium longicornis Latreille, 1804.
Eng.: Meadow Grasshopper; Dan.: Enggræshoppen; Fin.: Nurmiheinäsirkka, Sw.:
Kortvingad ängsgräshoppan.

Wide variations possible in coloration: green, yellow, brown and red. Side keels only
slightly introflexed. Transverse sulcus on pronotum only a little way posterior to the
mid-point, the prozona only being a few millimetres longer than the metazona. Fore
wings in male reach the tip of the abdomen or a bit less, the hind wings extending approximately to the middle of the fore wings (the hind wings can be seen through the
transparent fore wings). The fore wings in the female are flap-like and do not reach as
far as the middle of the abdomen. Tympanum one-third obscured. Valves of ovipositor
short. Numbers of stridulatory pegs: ♂ 70-130; ♀ 65-115. Length: ♂ 10-16 mm; ♀
16-23 mm.

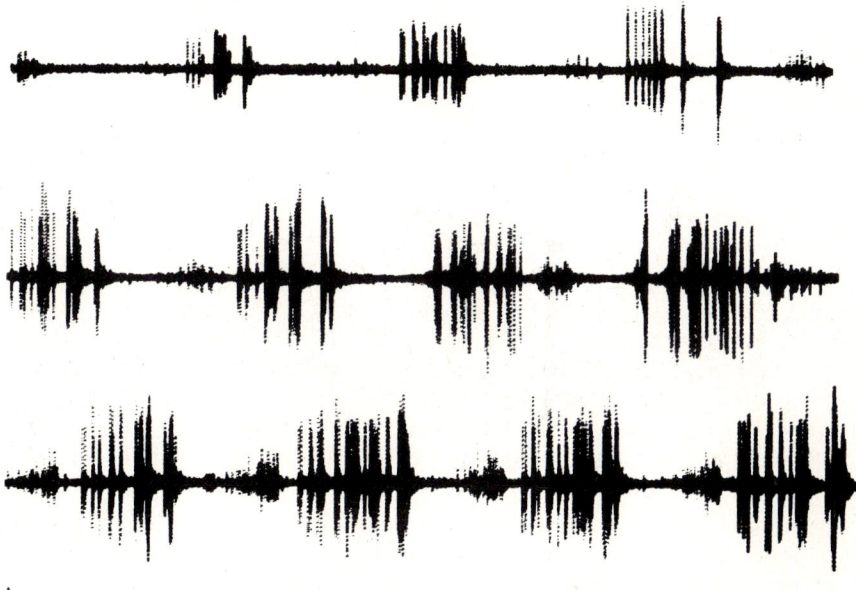

Fig. 86. Oscillogram of the song (one sequence with several chirps) of *Chorthippus parallelus*
(Zett.). — DK, NEZ: Tisvilde, August 1979. Scale: 0.5 sec.

Distribution. Commonly distributed in Schleswig-Holstein, Denmark and southern Sweden. Only found sporadically near the coast of southern Norway up to Stavanger. — It is common in southern Finland and the neighbouring areas of Russia. Occurs sporadically in northern Finland. Remarkably enough, it has not been found on the islands of Bornholm or Gotland, nor on other islands such as Samsø, Læsø or Anholt. This may indicate that it is a recent immigrant to the southerly portion of this area. — Scotland, England, Wales, France, Central Europe (mostly in mountains), Turkey, USSR and Mongolia.

Biology. Common in damp grass at the sides of ditches, at the foot of slopes, in meadows, and on the shores of lakes and beside streams.

The egg capsules are 5.1-13.2 mm in length, pear-shaped, terminating in a point superiorly. The surface of the capsule consists of hardened secretion and soil. It may contain up to ten eggs, diagonally positioned in relation to the longitudinal axis of the capsule. The eggs are 3.4-4.1 mm long, with a coarse hexagonal structure. The egg capsules are laid in the soil, just below ground level. Its development is reminiscent of that in other *Chorthippus* species. An extra nymphal instar has, however, been confirmed between the 2nd. and 3rd. stages.

Stridulation. The sequence lasts from one to two seconds and may be repeated at various rhythms. Each sequence consists of a series of chirps (about 4-5 a second) which merge to a certain extent but which can be distinguished as separate sounds. Each chirp consists of several syllables.

45. *Chorthippus (Ch.) montanus* (Charpentier, 1825)
    Fig. 87.

Gryllus montanus Charpentier, 1825, Hor. Ent.: 173.
Fin.: Nevaheinäsirkka; Sw.: Myrgräshoppan.

Highly reminiscent of the previous species in coloration and structure. The best identificatory feature is that the valves of the ovipositor are long. The numbers of stridulatory pegs in the file may be greater. Other special features include: — transverse sulcus of pronotum usually central, the prozona being the same size as the metazona; the fore wings in the male extend beyond the tip of the abdomen, the hind wings being long, extending to the outer quarter of the fore wings, and the fore wings in the female being rounded at the tip and covering about two thirds of the abdomen. Numbers of stridulatory pegs: ♂ 105-170; ♀ 105-165. Length: ♂ 10-16 mm; ♀ 16-22 mm.

Distribution. Holarctic. Northern Finland and Sweden and, in mountainous regions, further south in Sweden, e.g. to Härjedalen and Jämtland, in addition to Norway, where occasionally finds have been made in Hordaland and Telemarken. — It has also been found further south in Finland; round L. Ladoga in Russia, and *inter alia* in Skåne and Småland in Sweden. Schleswig-Holstein. — Sporadic in Central Europe and France, especially in mountainous regions; USSR; Mongolia.

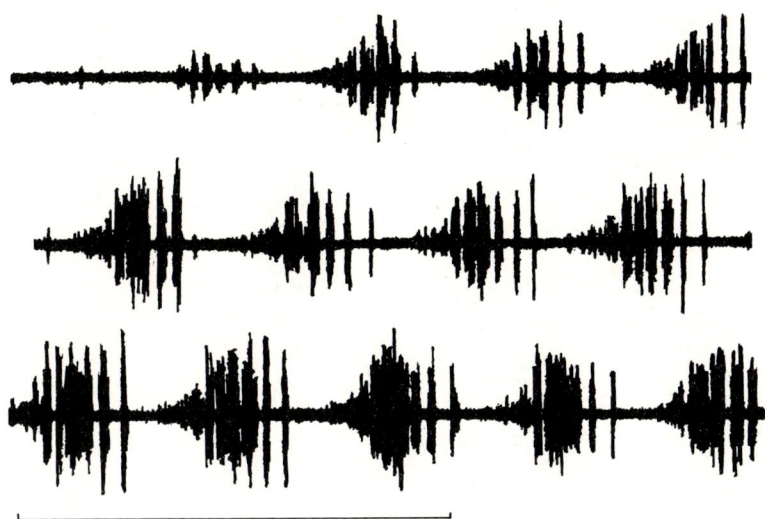

Fig. 87. Oscillogram of the song (one sequence with several chirps) of *Chorthippus montanus* (Charp.). — S, Upl.: Jumkil, August 1979. Scale: 0.5 sec. (From Wallin, 1979).

Biology. In damp places in bogs and in watery meadows with such plants as *Carex* and *Oxycoccus* (Lindberg, 1952).

Stridulation. The sequence lasts about three seconds and may be repeated at various rhythms. Each sequence consists of a series of chirps (about five a second), which are stronger, and more widely separated, than in the previous species. Each chirp consists of several syllables.

Genus *Myrmeleotettix* Bolivar, 1914

Myrmeleotettix Bolivar, 1914, Trab. Mus. Nat. Cienc. Nat., Ser. Zool. 20: 61.
Type-species: *Gomphocerus maculatus* Thunberg, 1815.

Foveolae narrow and rectangular. Antennae club-like in lateral view, most obvious in the male. Edge of fore wing straight with no widening at the base, the costal area not being much widened. Side keels on pronotum angularly introflexed in the prozona and highly divergent in the metazona. Tympanum two-thirds obscured. Valves of ovipositor non-denticulate.

Four species in Europe and Asia, of which only one in the area covered by the present survey.

46. *Myrmeleotettix m. maculatus* (Thunberg, 1815)
   Figs 50, 51A, 88.

Gomphocerus maculatus Thunberg, 1815, Mém. Acad. St. Pétersbg. 5: 221.
Eng.: Mottled Grasshopper; Dan.: Køllegræshoppen; Fin.: Nuijaheinäsirkka; Sw.: Klubbsprötgräshoppan.

Colour may vary widely: red, brown, green, yellow and black may be distributed in various areas. Antennae slightly longer than head and pronotum together. Fore wings and hind wings extend about as far back as the tip of the hind femora. Numbers of stridulatory pegs: ♂ 125-200; ♀ 115-160. Length: ♂ 12-13 mm; ♀ 14-16 mm.

   Distribution. Very widespread. In the west of Norway it gets up to about Sognefjord. The only parts of Sweden and Finland in which it is not found are the areas furthest north. — From the British Isles and France south to S.Italy and Greece; eastwards over the European part of the USSR to the Caucasus, Kazakhstan and Siberia. In S.Europe, mainly in the mountains.

   Biology. In sandy and dry places, and in "alvar" areas.

   Stridulation. The sequence lasts from 10 to 15 seconds, consisting of quite separate chirps (about two a second) which are initially weak but which finally attain full volume.

Fig. 88. Oscillogram of the song of *Myrmeleotettix maculatus* (Thnbg.). — S, Öl.: Vickleby, August 1978. Scale: 0.25 sec. (From Wallin, 1979).

Genus *Gomphocerus* Thunberg, 1815

Gomphocerus Thunberg, 1815, Mém. Acad. St. Pétersbg. 5: 221.
   Type-species: *Gryllus rufus* Linnaeus, 1758.

Foveolae narrow and rectangular. Antennae highly club-like, most markedly so in the male. Leading edge of fore wings extended slightly at the base, the costal area in the male being prominent, crowding the precostal area out and shortening it. Side keels on pronotum introflexed. Tympanum two-thirds obscured. Valves of ovipositor nondenticulate.
   Only one species.

47. *Gomphocerus rufus* (Linnaeus, 1758)
Figs 51B, 89, 90.

Gryllus rufus Linnaeus, 1758, Syst. Nat. ed. 10: 433.
Eng.: Rufous Grasshopper; Fin.: Iso nuijaheinäsirkka; Sw.: Stor klubbgräshoppan.

Basic coloration brownish. Antennae much longer than head and pronotum combined, dark, with prominent white tips. Fore wings and hind wings extend nearly to the end of the hind femora. Numbers of stridulatory pegs: ♂ 140-240; ♀ 130-200. Length: ♂ 14-18 mm; ♀ 17-24 mm.

Distribution. Not found in Denmark or Schleswig-Holstein. — In Sweden, it has been found from Skåne in the south to the most extreme northerly regions. In western Norway, it extends as far north as to Trondheim. — Not found in Finland, though it is present round L. Ladoga and on the Kola Peninsula in Russia. — South England, Wales, France, Central Europe, Italy, Yugoslavia, Hungary, Rumania, Bulgaria and USSR.

Biology. In open grass, e.g. at the edges of and in clearings in woods. Mesophil. Rare in damp and very dry localities.

Stridulation. The sequence lasts about four seconds. Individual chirps (about eight a second) merge to a certain extent but can be distinguished as separate sounds.

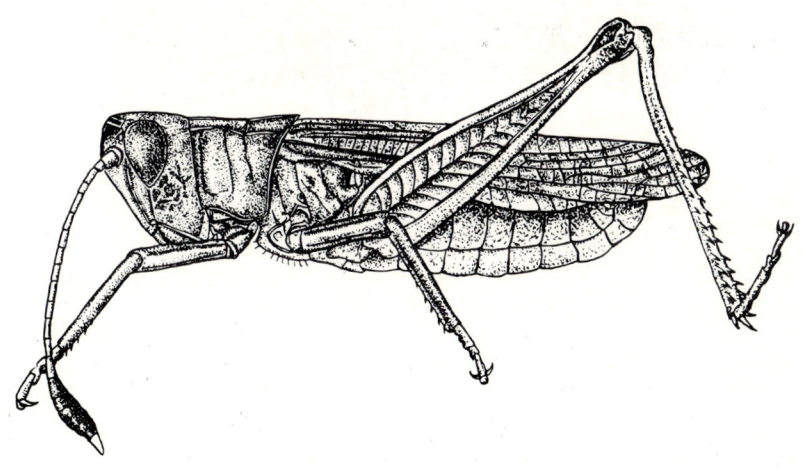

Fig. 89. *Gomphocerus rufus* (L.) ♂.
Length 14-18 mm. (From Harz, 1975).

115

Fig. 90. Oscillogram of the song of *Gomphocerus rufus* (L.). — S, Upl.: Järlåsa, August 1978. Scale: sec. (From Wallin, 1979).

Genus *Aeropedellus* Hebard, 1935

Aeropedellus Hebard, 1935, Ent. News Philad. 46: 184-188.
Type-species: *A. clavatus* (Thomson, 1873).

Foveolae narrow and rectangular. Antennae club-like, but in the female the widening-out is weak and may be entirely absent. Leading edge of fore wings slightly widened at base; costal area in male protuberant, crowding the precostal area out and thus shortening it. Side keels on pronotum angularly introflexed in the prozona and highly divergent in the metazona. Tympanum one-third obscured. Valves of ovipositor non-denticulate.

Five species in the Holarctic region. Only one in N.Europe.

48. *Aeropedellus variegatus* (Fischer-Waldheim, 1845)

Gomphocerus variegatus Fischer-Waldheim, 1845, Orth. Ross.: 341.
Fin.: Jänkäheinäsirkka.

Basic coloration brownish. Fore wings reach almost all the way down to the tip of the abdomen; hind wings slightly shorter. Antennae slightly shorter than head and pronotum combined. Length: ♂ 13-15 mm; ♀ 19-24 mm (Central European specimens).

Distribution. Boreo-alpine. In this area, only found in a few localities in northern Finland in the vicinity of Ivalo and Kemijärvi. The occurrences closest to here otherwise are in The Alps, the Apennines, Slovenia, Bulgaria, the Caucasus and Siberia.

Biology. South of Ivalo, in Finnish Lapland, at an altitude of 500 metres, it inhabits the lower reaches of the mountain moorland on gravelly soil where *Empetrum, Arctostaphylos alpina, Vaccinium vitis-idaea, Azalea procumbens* and clumps of *Festuca* were growing (Nordman, 1938).

References

Ahlén, I. & Degn, H. J. 1980. Lövvårtbitarens *Leptophyes punctatissima* sång. — Fauna och Flora 75: 265-268.

Albrecht, A. 1979. Utbredningen av rätvingar, kackerlaxkor och tvestjärtar i östra Fennoskandien (Orthoptera, Blattodea, Dermaptera). — Notulae ent. 59: 53-64.

Ander, K. 1945. Catalogus Insectorum Sueciae. V. Orthoptera. — Opusc. ent. 10: 127-134.

– 1947. Flygförmågan hos våra hopprätvingar. — Fauna och Flora 42: 210-221.

– 1949a. Die boreoalpinen Orthopteren Europas. — Opusc. ent. 14: 89-105.

– 1949b. *Omocestus haemorrhoidalis* Charp. in Sweden. — Ibid. 14: 121-149.

– 1949c. Rassenbildung und Variabilität bei der skandinavischen *Platycleis denticulata* Panz. (Salt. Tettig.). — Kungl. Fysio. Sällsk. Lund Förh. 19 (3): 1-24.

– 1953. Catalogus Insectorum Sueciae. V. Orthoptera (Dermaptera, Blattoidea, Saltatoria). — Opusc. ent. 18: 88.

Bei-Bienko, G. Y. & Mishchenko, L. L. 1951. Locusts and grasshoppers of the U.S.S.R. and adjacent countries. I and II. — Israel Program Scient. Transl. (English edition 1963-64).

Blunck, H. 1942. *Leptophyes punctatissima* Bosc als Rosenschädling. — Planzenkrankh. 52: 192-204.

Chopard, L. 1951. Orthoptéroides. — Faune de France 56: 359 pp., Paris.

Elsner, N. 1974. Neuroethology of sound production in gomphocerine grasshoppers (Orthoptera: Acrididae). I. Song pattern and stridulatory movements. — J. comp. Physiol. 88: 67-102.

Faber, A. 1953. Laut- und Gebärdensprache bei Insekten: Orthoptera (Geradflügler). Teil 1. — Mitt. st. Mus. Naturk. Stuttg. 287: 198 pp.

Fabricius, J. C. 1775. Systema entomologiae. 832 pp., Flensburgi et Lipsiae.

Farrow, R. A. 1965. The post-embryonic development of the external genitalia of *Tetrix* Latreille (Orthoptera: Tetrigidae). — Ann. Mag. nat. Hist. (13) 7: 301-313.

Hartley, J. C. 1964. The structure of the eggs of the British Tettigoniidae (Orthoptera). — Proc. R. ent. Soc. Lond. (A) 39: 111-117.

Hartley, J. C. & Warne, A. C. 1973. The distribution of *Pholidoptera griseoaptera* (De Geer) (Orthoptera: Tettigoniidae) in England and Wales related to accumulated temperatures. — J. Anim. Ecol. 42: 531-537.

Harz, K. 1957. Die Geradflügler Mitteleuropas. 494 pp., Jena.

– 1969. Die Orthopteren Europas. The Orthoptera of Europe. Vol. I. — Series Entomologica 5: 749 pp., The Hague.

– 1975. Die Orthopteren Europas. The Orthoptera of Europe. Vol. II. — Series Entomologica 11: 939 pp., The Hague.

117

Hellén, W. 1921. Veränderungen in der Insektenfauna Finnlands bis zum Jahr 1921. Dermaptera, Orthoptera, Blattaria. — Notulae ent. 1: 60-61.

Holst, K. T. 1965. Om *Conocephalus dorsalis* Latreille (Sivgræshoppen) i Danmark. — Flora og Fauna 71: 61-64.

– 1969. The distribution of Orthoptera in Denmark, Scania and Schleswig-Holstein. — Ent. Meddr 37: 413-442.

– 1970. Kakerlakker, græshopper og ørentviste. — Danmarks Fauna 79: 221 pp., København.

– 1976. Sjælden græshoppe *(Chorthippus mollis)* fundet almindelig på Samsø. — Flora og Fauna 82: 85.

Jago, N. D. 1971. A review of the Gomphocerinae of the World with a key to the genera. — Proc. Acad. nat. Sci. Philad. 123 (8): 122-343.

Jensen, P. 1966. *Chorthippus vagans* Eversm. Ny markgræshoppe for den danske fauna. — Flora og Fauna 72: 34-36.

Johnsen, P. 1976. Markgræshoppen *Oedipoda caerulescens* genfundet i Danmark. — Ibid. 82: 70.

Jones, M. D. R. 1966. The acoustic behaviour of the Bush Cricket *Pholidoptera griseoaptera*. — J. exp. Biol. 45: 15-44.

Klingstedt, H. 1933. Die Verteilung der Orthopteren der Schären W-Nylands auf Längszonen- — Notulae ent. 13: 61-62.

Knaben, N. 1943. Oversikt over Norges Orthoptera. — Bergens Museums Årbok 2: 4-43.

Larsen, A. 1944. Markfaarekyllingen *(Gryllus campestris)* paa Bornholm. — Flora og Fauna 50: 1-3.

Lindberg, H. & Saris, N.-E. 1952. Insektfaunan i Pisavaara Naturpark. — Acta Soc. Fauna Flora Fennica 69 (2): 8-9.

Linnaeus, C. 1758. Systema naturae per regni tria naturae. Ed. 10, vol. 1: 824 pp., Holmiae.

– 1961. Fauna svecica sistens animalia Sveciae regni. Ed. 2, 578 pp., Stockholmiae.

– 1767. Systema naturae per regni tria naturae. Ed. 12 (rev.). Holmiae.

Lunau, C. 1950. Zur Heuschreckenfauna Schleswig-Holsteins. — Schr. naturw. Ver. Schlesw.-Holst. 24: 51-56.

– 1972. *Chorthippus vagans* Eversm., eine auch in Schleswig-Holstein gefundene Art. — Bombus 2: 201.

Miram, E. 1931. Beitrag zur Kenntnis der Orthopterenfauna der nördlichen Polarzone mit Berücksichtigung der Dermaptera und Blattodeen. — Zool. Anz. 97: 37-46.

Nielsen, E. T. 1938. Zur Ökologie der Laubheuschrecken. — Ent. Meddr 20: 121-164.

Nielsen, E. T. & Dreisig, H. 1970. The behavior of stridulation in Orthoptera Ensifera. — Behavior 37 (3-4): 205-252.

Nordman, A. 1938. Om förekomsten av gräshoppsarten *Gomphocerus variegatus* Fish.-Waldh. in N-Finland. — Memoranda Soc. Fauna Flora Fennica 14: 124-125.

- 1963a. Die in Finnland vorkommende Sandschrecke, *Sphingonotus coerulans* (L.) *cyanopterus* Charp., ist jedenfalls kein guter Flieger. — Ibid. 39: 141-143.

- 1963b. Zur Kenntnis der Orthopterenfauna auf den isolierten Ausseninseln mitten im Finnischen Meerbusen. — Ibid. 39: 158-162.

Nørgaard, E. 1942. Bidrag til hedeskrattens biologi *(Bryodema tuberculata* F.). — Flora og Fauna 48: 1-17.

Panelius, S., 1978. The detailed geographical distribution of *Tettigonia cantans* in Finland (Orthoptera, Tettigoniidae). — Notulae ent. 58: 151-157.

Petersen, E. 1909. Ørentviste, Kakerlakker og Græshopper. — Danmarks Fauna 6: 41 pp., København.

Ragge, D. R. 1965. Grasshoppers, crickets and cockroaches of the British Isles. 299 pp., London & New York.

Richards, O. W. & Waloff, N. 1954. Studies on the biology and population dynamics of British grasshoppers. — Anti-Locust Bull. 17: 182 pp., London.

Richards, T. J. 1958. Observations on the nymphs of seven Tettigonioids. — Entomologist 91: 53-66.

Sandhall, Å. & Ander, K. 1978. Gräshoppor, syrsor och deras släktingar. 93 pp., ICA bokförlag, Västerås.

Valle, K. J. 1930. Die Orthopterenfauna der nördlichsten Teile von Ostfennoskandia mit besonderer Berücksichtigung des Petsamo-Gebiets. — Notulae ent. 10: 40-42.

Wallin, L. 1979. Svenska gräshoppors og vårtbitares sångläten. Texthäfte til ljudband. — Zoologiska museet, Uppsala.

Warne, A. C. & Hartley, J. C. 1975. The distribution and dispersal of *Conocephalus dorsalis* (Latreille) (Tettigoniidae) in the British Isles. — Entomologist's Gaz. 26: 127-132.

Weidner, H. 1950 & 1951. Beitrag zur Geradflüglerfauna Schleswig-Holsteins. — Mitt. faun. ArbGemein. Schleswig-Holstein, Hamburg u. Lübeck 3: 15-17, 4: 12-14.

		DENMARK												
		Schl.-Holst.	G. Britain	SJ	EJ	WJ	NWJ	NEJ	F	LFM	SZ	NWZ	NEZ	B	Sk.	Bl.

		Schl.-Holst.	G. Britain	SJ	EJ	WJ	NWJ	NEJ	F	LFM	SZ	NWZ	NEZ	B	Sk.	Bl.
Leptophyes punctatissima (Bosc)	1	●	●	●	●		●	●	●	●			●	●	●	
Meconema thalassinum (De Geer)	2	●	●	●	●			●	●	●	●	●	●	●	●	●
Conocephalus dorsalis (Latr.)	3	●	●	●	●	●		●	●	●	●	●	●	●	●	●
Tettigonia viridissima (L.)	4	●	●	●	●		●	●	●	●	●	●	●	●	●	●
T. cantans (Fuessly)	5	●		●	●				●							
Decticus verrucivorus (L.)	6	●	●	●	●	●	●	●	●	●	●	●	●	●	●	●
Platycleis albopunctata (Gz.)	7	●	●		●								●	●	●	
Metrioptera bicolor (Phil.)	8	●													●	
M. roeseli (Hag.)	9	●	●			●			●							
M. brachyptera (L.)	10	●	●	●	●	●	●	●				●		●	●	●
Pholidoptera griseoaptera (De Geer)	11	●	●	●	●			●	●	●	●	●	●	●	●	●
Tachycines asynamorus Ad.	12	i	n	t	r	o	d	u	c	e	d					
Gryllus campestris L.	13	●	●										●			
Acheta domestica (L.)	14	i	n	t	r	o	d	u	c	e	d					
Gryllotalpa gryllotalpa (L.)	15	●	●		●				●	●	●	●	●	●	●	●
Tetrix subulata (L.)	16	●	●	●	●	●	●		●	●	●	●	●	●	●	●
T. fuliginosa (Zett.)	17															
T. undulata (Sow.)	18	●	●	●	●	●	●	●	●	●	●	●	●		●	●
T. bipunctata (L.)	19			●			●							●	●	●
T. nutans (Hag.)	20	●														
Podisma pedestris (L.)	21															
Melanoplus frigidus (Boh.)	22															
Psophus stridulus (L.)	23														●	●
Locusta migratoria L.	24															
Oedipoda caerulescens (L.)	25	●		●									●			
Bryodema tuberculata (F.)	26	●			●	●	●	●								
Sphingonotus caerulans (L.)	27	●														
Mecostethus grossus (L.)	28	●	●	●	●	●	●	●	●		●	●	●	●	●	●
Chrysochraon dispar (Germ.)	29	●														
Stenobothrus lineatus (Panz.)	30	●	●													
S. stigmaticus (Ramb.)	31	●	●													
Omocestus viridulus (L.)	32	●	●	●	●	●	●	●	●	●	●	●	●	●	●	●
O. ventralis (Zett.)	33		●												●	●
O. haemorrhoidalis (Charp.)	34	●			●			●							●	
Stauroderus scalaris (F.-W.)	35															
Chorthippus brunneus (Thnbg.)	36	●	●	●	●	●	●	●	●	●	●	●	●	●	●	●
C. mollis (Charp.)	37	●		●	●											
C. biguttulus (L.)	38	●		●	●	●		●	●	●	●	●	●	●	●	●

120

	Hall.	Sm.	Öl.	Gtl.	G. Sand.	Ög.	Vg.	Boh.	Dls.	Nrk.	Sdm.	Upl.	Vstm.	Vrm.	Dlr.	Gstr.	Hls.	Med.	Hrj.	Jmt.	Ång.	Vb.	Nb.	Ås. Lpm.	Ly. Lpm.	P. Lpm.	Lu. Lpm.	T. Lpm.
1	●		●	●		●	●																					
2	●		●			●	●																					
3	●	●	●	●		●	●					●																
4	●	●	●	●	●	●	●	●	●	●	●	●	●															
5																												
6	●	●	●	●		●	●	●	●	●	●	●	●	●	●	●				●	●				●			
7	●	●	●	●	●	●		●																				
8																												
9												●																
10	●	●	●	●		●	●	●	●	●	●	●	●	●	●	●				●	●	●	●				●	●
11	●	●	●	●	●	●	●	●			●	●		●														
12																												
13																												
14																												
15	●	●	●	●			●																					
16	●	●	●	●		●	●	●	●	●	●	●	●	●	●	●				●	●	●	●		●			
17														●		●				●		●	●				●	●
18	●	●	●	●		●	●	●	●	●	●	●	●	●	●	●				●	●	●						●
19	●	●	●	●		●	●	●	●	●	●	●	●	●	●	●				●	●	●	●		●			
20																												
21		●	●	●		●	●	●	●	●	●	●	●	●	●	●				●	●	●	●	●			●	
22																		●	●					●	●	●	●	
23		●	●	●		●	●	●	●				●	●														
24																												
25	●																											
26			●																									
27			●	●		●		●																				
28	●	●	●	●		●	●		●	●	●	●		●	●	●	●			●	●	●	●	●				●
29		●	●									●										●						
30																												
31																												
32	●	●	●	●		●	●	●	●	●	●	●	●	●	●	●	●	●		●	●				●			
33	●	●	●			●	●		●		●																	
34			●	●		●																						
35			●																									
36	●	●	●	●		●	●	●	●	●	●	●	●	●							●	●			●			
37																												
38	●	●	●			●	●	●	●	●	●			●	●					●	●	●						

		Ø+AK	HE (s+n)	O (s+n)	B (ø+v)	VE	TE (y+i)	AA (y+i)	VA (y+i)	R (y+i)	HO (y+i)	SF (y+i)	MR (y+i)	ST (y+i)	NT (y+i)	Ns (y+i)
Leptophyes punctatissima (Bosc)	1	●				●	●	●	●							
Meconema thalassinum (De Geer)	2				●				●							
Conocephalus dorsalis (Latr.)	3	●			●											
Tettigonia viridissima (L.)	4	●	●		●	●	●	●	●	●						
T. cantans (Fuessly)	5															
Decticus verrucivorus (L.)	6	●	●	●	●	●	●	●	●	●	●					
Platycleis albopunctata (Gz.)	7	●							●							
Metrioptera bicolor (Phil.)	8															
M. roeseli (Hag.)	9															
M. brachyptera (L.)	10	●		●			●	●	●	●						
Pholidoptera griseoaptera (De Geer)	11	●			●	●										
Tachycines asynamorus Ad.	12															
Gryllus campestris L.	13															
Acheta domestica (L.)	14															
Gryllotalpa gryllotalpa (L.)	15															
Tetrix subulata (L.)	16	●	●	●	●	●	●	●			●	●		●		
T. fuliginosa (Zett.)	17															
T. undulata (Sow.)	18	●		●	●	●	●			●	●	●	●			
T. bipunctata (L.)	19	●	●	●	●	●	●	●		●	●	●	●	●		●
T. nutans (Hag.)	20															
Podisma pedestris (L.)	21	●	●	●	●		●	●	●	●	●	●				●
Melanoplus frigidus (Boh.)	22			●	●						●	●	●	●	●	●
Psophus stridulus (L.)	23	●	●		●	●	●									
Locusta migratoria L.	24															
Oedipoda caerulescens (L.)	25															
Bryodema tuberculata (F.)	26															
Sphingonotus caerulans (L.)	27	●			●	●	●									
Mecostethus grossus (L.)	28	●			●	●	●	●					●	●		
Chrysochraon dispar (Germ.)	29															
Stenobothrus lineatus (Panz.)	30															
S. stigmaticus (Ramb.)	31															
Omocestus viridulus (L.)	32	●	●	●	●	●	●	●	●	●	●	●	●	●	●	
O. ventralis (Zett.)	33	●		●			●	●			●					
O. haemorrhoidalis (Charp.)	34															
Stauroderus scalaris (F.-W.)	35															
Chorthippus brunneus (Thnbg.)	36	●		●		●	●	●	●	●	●	●			●	
C. mollis (Charp.)	37															
C. biguttulus (L.)	38	●			●	●	●									

	Nn (ø+v)	TR (y+i)	F (v+i)	F (n+ø)	Al	Ab	N	Ka	St	Ta	Sa	Oa	Tb	Sb	Kb	Om	Ok	ObS	ObN	Ks	LkW	LkE	Le	Li	Vib	Kr	Lr
1																											
2																											
3					●	●	●	●	●																●		
4					●	●	●																				
5									●			●	●	●	●										●	●	
6					●	●	●	●	●	●	●	●	●	●	●	●	●								●	●	●
7					●																						
8																											
9					●	●	●	●	●	●	●	●	●	●	●	●	●								●	●	
10					●	●	●	●	●	●	●	●	●	●	●	●	●	●	●	●					●	●	
11					●	●	●	●	●																●		
12																											
13																											
14																											
15																											
16			●		●	●	●	●	●	●	●	●	●	●	●	●	●	●	●	●					●	●	●
17														●	●	●	●	●	●				●	●		●	●
18					●	●																					
19		●	●	●	●	●	●	●	●	●	●	●	●	●	●	●	●	●	●	●	●	●	●	●	●	●	●
20						●	●				●		●	●											●	●	
21		●	●	●	●	●	●	●	●	●	●	●	●	●	●	●	●	●	●	●	●	●	●	●	●	●	
22	●	●	●	●										●			●		●	●			●	●		●	
23						●	●	●	●	●	●	●	●	●	●	●	●								●	●	
24																											
25																											
26																									●	●	
27						●	●																		●		
28					●	●	●	●	●	●	●	●	●	●	●	●	●	●	●	●					●	●	
29					●	●	●	●	●			●		●				●	●						●		
30																											
31																											
32					●	●	●	●	●	●	●	●	●	●	●								●		●	●	
33														●													
34																											
35																											
36					●	●	●	●	●	●	●	●	●	●	●	●									●	●	●
37																									●		
38					●	●	●	●	●	●	●		●	●	●		●	●									

DENMARK

		Schl.-Holst.	G. Britain	SJ	EJ	WJ	NWJ	NEJ	F	LFM	SZ	NWZ	NEZ	B	Sk.	Bl.
C. apricarius (L.)	39	●		●	●				●	●	●			●	●	●
C. vagans (v.)	40	●	●					●								
C. pullus (Phil.)	41															
C. albomarginatus (De Geer)	42	●	●	●	●	●	●	●	●	●	●	●	●	●	●	●
C. dorsatus (Zett.)	43	●		●	●	●		●	●	●	●	●	●		●	●
C. parallelus (Zett.)	44	●	●	●	●	●	●	●	●	●	●	●	●		●	●
C. montanus (Charp.)	45	●													●	
Myrmeleotettix maculatus (Thnbg.)	46	●	●	●	●	●	●	●	●	●		●	●	●	●	●
Gomphocerus rufus (L.)	47		●												●	●
Aeropedellus variegatus (F.-W.)	48															

NORWAY

		Ø+AK	HE (s+n)	O (s+n)	B (ø+v)	VE	TE (y+i)	AA (y+i)	VA (y+i)	R (y+i)	HO (y+i)	SF (y+i)	MR (y+i)	ST (y+i)	NT (y+i)	Ns (y+i)
C. apricarius (L.)	39															
C. vagans (v.)	40															
C. pullus (Phil.)	41															
C. albomarginatus (De Geer)	42	●			●	●		●	●	●	●					
C. dorsatus (Zett.)	43															
C. parallelus (Zett.)	44	●			●	●	●			●						
C. montanus (Charp.)	45						●				●					
Myrmeleotettix maculatus (Thnbg.)	46	●	●	●	●	●	●	●	●	●	●					
Gomphocerus rufus (L.)	47	●	●	●	●	●	●	●	●	●	●			●		
Aeropedellus variegatus (F.-W.)	48															

	Hall.	Sm.	Öl.	Gtl.	G. Sand.	Ög.	Vg.	Boh.	Dlsl.	Nrk.	Sdm.	Upl.	Vstm.	Vrm.	Dlr.	Gstr.	Hls.	Med.	Hrj.	Jmt.	Ång.	Vb.	Nb.	Ås. Lpm.	Ly. Lpm.	P. Lpm.	Lu. Lpm.	T. Lpm.
39	●	●	●	●																								
40																												
41																												
42	●	●	●	●		●	●	●	●	●	●	●		●	●	●												
43		●	●			●		●		●																		
44	●	●	●			●	●	●	●		●	●																
45		●				●	●	●	●						●	●		●	●		●	●		●				
46	●	●	●	●	●	●	●	●	●	●	●	●		●	●	●	●	●		●	●	●		●				
47	●	●	●	●		●	●	●	●	●	●	●	●	●	●	●	●	●		●	●	●	●	●	●	●	●	
48																												

	Nn (ø+v)	TR (y+i)	F (v+i)	F (n+ø)	Al	Ab	N	Ka	St	Ta	Sa	Öa	Tb	Sb	Kb	Om	Ok	Ob S	Ob N	Ks	LkW	LkE	Le	Li	Vib	Kr	Lr
39																											
40																											
41																									●	●	
42					●	●	●		●	●	●	●				●		●	●						●	●	
43																											
44					●	●	●	●	●	●	●		●	●	●		●	●	●			●			●	●	
45						●	●		●		●	●	●	●	●	●	●	●	●						●	●	●
46					●	●	●	●	●	●	●	●	●	●	●	●	●	●	●	●					●	●	●
47																									●	●	●
48																			●					●			

125

Index

Reference is given to the keys and the main treatment. Synonyms are in italics.

126